THE JOY OF BIRDWATCHING

THE JOY OF BIRDWATCHING

ALAN DAVIES & RUTH MILLER

summersdale

THE JOY OF BIRDWATCHING

Copyright © Summersdale Publishers Ltd, 2015

Illustrations © Shutterstock

All rights reserved.

No part of this book may be reproduced by any means, nor transmitted, nor translated into a machine language, without the written permission of the publishers.

Alan Davies and Ruth Miller have asserted their right to be identified as the author of this work in accordance with sections 77 and 78 of the Copyright, Designs and Patents Act 1988.

Condition of Sale
This book is sold subject to the condition that it shall not, by way of trade or otherwise, be lent, re-sold, hired out or otherwise circulated in any form of binding or cover other than that in which it is published and without a similar condition including this condition being imposed on the subsequent purchaser.

Summersdale Publishers Ltd
46 West Street
Chichester
West Sussex
PO19 1RP
UK

www.summersdale.com

Printed and bound in the Czech Republic

ISBN: 978-1-84953-670-7

Substantial discounts on bulk quantities of Summersdale books are available to corporations, professional associations and other organisations. For details contact Nicky Douglas by telephone: +44 (0) 1243 756902, fax: +44 (0) 1243 786300 or email: nicky@summersdale.com.

Acknowledgements

Ruth: I'd like to thank my parents, Don and Cecile Miller, for teaching me the birds of rural Kent as I grew up, and for planting the seed which grew to a birding passion in later years.

Alan: I'd like to thank my parents, Peter and Patricia Davies, for sparking my interest in wild birds at an early age. I doubt they realised they were starting a lifelong obsession!

About the Authors

Alan Davies and **Ruth Miller** met through their mutual work at The Royal Society for the Protection of Birds (RSPB). Having been a keen birder for 30 years, Alan was the manager of the Society's flagship nature reserve at RSPB Conwy on the North Wales coast, while Ruth was the Head of Trading, setting up the RSPB's own operation to source and supply birdcare.

In 2012, they embarked on a year-long adventure, travelling the world aiming to set a new world record for the highest number of bird species seen in a single calendar year. On their epic quest, they visited 27 countries and saw a world-record-breaking 4,341 bird species. On returning to Britain, they wrote a book telling the tale: *The Biggest Twitch*.

Now based back in North Wales in the United Kingdom, Alan and Ruth still continue to travel, but this time sharing their passion for birds with others. They run a bird tour company called 'Birdwatching Trips with the Biggest Twitch', offering guided birdwatching tours in the best birding locations across the British Isles and in Europe. When they're not out exploring the best locations for wildlife, they're writing about it, with a published guide to the *Best Birdwatching Sites in North Wales* and a regular blog on their own website www.birdwatchingtrips.co.uk covering their continuing birding adventures around Britain and the rest of the world.

CONTENTS

INTRODUCTION

*Birding is hunting without killing,
preying without punishing, and
collecting without clogging your home.*
MARK OBMASCIK

People sometimes ask us, 'What do you do when you're not birdwatching?' The answer is, 'We're always birdwatching!' This isn't because we're absolute birdwatching fanatics, although some people might say we are. It is because we can always watch birds.

Unlike many leisure activities, you don't have to go to a special venue or live in a particular part of the country to go birdwatching – you can see birds anywhere. It isn't restricted to specific times of day or certain months of the year – you can watch birds all year round at any time of day or night. And you don't need any particular equipment either – all you need is a pair of eyes, a pair of ears, and the ability to look and listen. A pair of binoculars will help you get a better view and a field guide will tell you more about what particular bird you are looking at, but the real pleasure comes from simply seeing birds. Regardless of whether you can put a name to them or not, birds are beautiful, fascinating creatures and they're everywhere.

Even while we're writing this, we can see a Collared Dove sitting on the chimneypots, a Buzzard is circling over the hills in the distance, a Jackdaw is probing at something it has found on the pavement opposite and a Herring Gull has just landed on a nearby rooftop. That's four birds we're watching, just while we're working on the computer.

It's easy to start birdwatching and anyone can do it. All you need to do is look, and notice. And once you start looking and noticing one bird, you'll start looking and noticing more. You'll notice the differences in their colour; maybe they hop or they strut along the ground; perhaps they'll be soaring on the wind or hovering to keep still.

7

Before you know it, you'll be looking out for birds. You'll be looking to see if there are as many Pied Wagtails marching around the supermarket car park as last time you did the shopping. Or you'll be counting the number of Kestrels you spot from the motorway. Or how many times the Robin comes into your garden to feed just as you're making your first coffee of the day.

And once you start looking out for birds, you'll notice something else too. You'll notice that you've forgotten whatever it was you were worrying about; that important meeting suddenly isn't so daunting, or perhaps your tired feet don't ache any more. Because birdwatching is good for you. While you're concentrating on seeing and enjoying birds, you're not thinking about anything else. It's absorbing without being taxing, and stimulating without wearing you out. If you've just watched a Blackbird foraging for worms amongst the leaf litter, you can't help but smile inside and your day is already better for it.

KEY PEOPLE IN BIRDING HISTORY

The love for all living creatures is the most noble attribute of man.

CHARLES DARWIN

Birdwatching hasn't always been the specific leisure activity in Britain that it is today. In earlier centuries, when more people lived in the country and made their living on the land, birds and other wildlife simply inhabited the same space, side by side with humans, and went largely unobserved in any formal sense. Only a relatively small number of people actively watched wildlife and a disproportionately large number of these were members of the clergy. Perhaps they, more than many other people, had the opportunity and the means to explore the countryside as they visited their parish.

Of these, **Gilbert White** (1720–1793) is perhaps the best known and often considered one of our earliest birdwatchers and ornithologists. Born in a vicarage in Selborne in Hampshire, he was ordained in 1749 and was the curate of a number of parishes across the Shires, including Selborne itself. He is best known for his book

The Natural History and Antiquities of Selborne, written in 1789. Based on a series of letters and notes, it chronicles his observations of birds, mammals, insects, plants and other wildlife in Selborne in a detailed and descriptive manner. Not afraid to challenge the scientific beliefs of the time, he employed a group of men to dredge his local ponds to disprove the theory that Swallows spent the winter buried in mud at the bottom of pools and lakes. Of course, no matter how deep they dug, they weren't able to uncover a single Swallow. To further prove his case, one season he tied lengths of coloured cotton to the legs of the Swallows nesting around his house and was delighted to see the tagged birds returning to the same site the following spring. Perhaps this was the earliest recorded example of bird ringing in Britain?

Nearly a century later, another naturalist also challenged the status quo of scientific thinking. **Charles Darwin** (1809–1882) had a unique opportunity to study a variety of birds, plants and fossils first-hand when he joined a 5-year voyage around the world on the HMS *Beagle* in 1831 as the resident naturalist. He studied examples of Mockingbirds collected from different islands across the Galapagos archipelago and noticed that, while they seemed superficially to be the same species, there were slight variations amongst them, such as the length of their bills, depending on which island they lived on. He suggested that this adaptation might have developed over time to allow them to capitalise on the particular circumstances on each separate island. This became his revolutionary theory on the origin of species. Darwin's theory of evolution or 'survival of the fittest' (though maybe 'survival of the most adaptable' is more accurate) rocked the scientific world. It was radical stuff, as it challenged the accepted teachings from the Bible that in the beginning God created the world just as they found it.

But although he gets most of the credit, Darwin wasn't the only birdwatcher to come up with the idea of evolution through natural selection. Remarkably, **Alfred Russel Wallace** (1823–1913), a contemporary of Charles Darwin, but working completely independently, also developed the theory that species evolved over time to best suit their surroundings, based on his own fieldwork and studies in the Amazon Basin and Indonesia. Both Wallace and Darwin published papers on the subject, though Darwin has always had the wider acknowledgement. Perhaps groundbreaking theories are like buses; none for ages and then two come along at once.

Meanwhile, over in the United States, a gentleman called **John James Audubon** (1785–1851) was making his mark on the world of birdwatching through his magnificent bird paintings. A passionate ornithologist and talented artist, he spent much of his life in rural America hunting and drawing birds, and discovered 25 new bird species as he did so. Unlike earlier artists, he painted birds realistically against the backdrop of their typical habitats. His most important work was *Birds of America*, which he sold to patrons in both America and Britain through subscriptions in advance, perhaps the original collector's part-work. In order to make his illustrations as accurate as possible, Audubon shot many of his specimens, as this was the only way to get a close enough look at the birds since binoculars were not available in those days. He painted his subjects life-size, which meant that even though he was painting on paper in Double Elephant size – each sheet was a massive 39 x 25 inches (99 x 64 cm) – he had to contort some of the larger birds into the most peculiar positions in order to fit them onto the page. For example, his Great Blue Heron is almost touching its toes with its bill to get it all on one page. Nonetheless, his birds have a bewitching liveliness to them and his paintings are still highly sought-after today, though you would need a very large bookcase and a lot of money to own an original edition: a copy of *Birds of America* reached nearly £7m at auction in 2010.

Fast-forward to the twentieth century and Audubon's influence was still being felt in the illustrations of birds in field guides. However, a quantum leap in the style of these field guides was made by another American birdwatcher and artist, **Roger Tory Peterson** (1908–1996). His eureka moment was to introduce the technique of lines pointing to the key field marks on the illustration, to show what particular features to concentrate on when identifying each bird. Like all great ideas, its beauty lies in its simplicity and Peterson is credited with creating the first modern field guide.

Back in Britain, another prominent ornithologist and artist was promoting birdwatching on the ground. **Sir Peter Markham Scott** (1909–1989) had grown up as a passionate supporter of birds and wildlife. Those famous instructions from his father, Robert Scott (the ill-fated Scott of the Antarctic), to 'make the boy interested in natural history' had paid dividends and Sir Peter Scott was extremely talented in painting birds and other wildlife, as well as creating the highly effective camouflage design used on naval vessels in World War Two. In 1946 he founded the Severn Wildfowl Trust (now known as the Wildfowl & Wetlands Trust, or WWT) at Slimbridge in Gloucestershire, and by instigating a captive breeding programme he saved the Nene Goose from extinction. The WWT now has nine reserves across the country welcoming over a million visitors a year, all coming to enjoy an encounter with birds.

And how many of us have been absorbed as **Sir David Attenborough** (1926–present) brought birds into our homes through his wonderful television programmes about wildlife over the years. In 1998 a new series, *The Life of Birds* was first broadcast. It had been three years in the making, with footage shot in 42 countries, and it covered in fascinating detail how birds evolved and all aspects of how birds live out their daily lives. It brought some of the most exotic bird species in the world right into our living rooms.

These people, among many other advocates and enthusiasts, are part of the rich heritage of modern birdwatching, which is continued today by amateurs and experts alike. The more we learn ourselves, the more we realise there is still to learn, and today's emerging birdwatchers may be tomorrow's pioneers in the birding world.

CHAPTER 2

BINS AND BOOKS

*Before I came along, the primary way to observe
birds was to shoot them and stuff them.*

ROGER TORY PETERSON

Birdwatching is an easy-access activity, and you can start to enjoy watching birds without having to buy lots of expensive equipment.

Choosing Your Perfect Bins!

While some people like to practise 'naked birdwatching' (that's without binoculars, not clothes!), most birdwatchers would agree that you really need a pair of binoculars to get the most out of watching birds. Quick and easy to use, binoculars will allow you to see birds in much more detail than the naked eye, whether they're coming to your garden feeder, pottering about on an estuary or flying overhead. Binoculars, often called 'bins' for short, come in all sorts of makes, types and prices and at first glance there may be a bewildering amount of choice. However, it is possible to buy the perfect pair of binoculars to start your birdwatching without spending a huge amount of money.

Magnification

Choosing the right binoculars is all about numbers – for example, you're likely to encounter figures such as 8x32, 10x42 or 12x50. What does it all mean? The sets of numbers all boil down to two things: the magnification and the size of the lenses. The first number refers to how many times the image is magnified, e.g. eight times, ten times or even twelve times, but more magnification isn't necessarily better. A lower magnification such as eight times will give you a wider field of view; in other words, you will have a wider landscape in focus through your bins, making it easier to spot birds. Twelve times magnification will give you an image that is larger, but of a narrower view, and bear in mind that any unsteady movements of the hands while using the bins will be magnified twelve times too. In general, lower magnification means the image is brighter, the field of view is wider and the binoculars are easier to hold steady.

Lens Diameter

The second number refers to the diameter (in millimetres) of the larger lens at the far end of your bins, called the objective lens. A larger diameter will let in more light, making it easier to see birds in low-light conditions such as at dawn or dusk, or in dense woodland. However, the larger the lens the heavier it will be, so you may find the trade-off in greater light-gathering capability makes a larger lens too heavy to use comfortably. And there's no point in having a pair of binoculars that would be great to use for your dawn birdwatching if only you could lift them!

If you are likely to be watching birds while out walking, then weight will be a consideration and a pair of 8x32s or 10x32s would be ideal. If you will be spending more time sitting in a hide or scanning out to sea or across wide estuaries, then a pair of 10x42s may be better for you. Of course, a higher-quality pair of 8x32s will still give you a brighter image than a cheap pair of 10x42s, but ultimately it all comes down to personal choice: what feels right in your hands, what you like looking through and how much money you want to spend. It is essential that you have a real hands-on test, so make sure you try before you buy to get what's right for you. To give you an example,

Ruth uses 8x32s whereas Alan, who has larger hands, uses 8x42s, as we each find that particular model works best for us.

Amazing Bird Fact

Peregrine Falcons can reach speeds of up to 200 mph (322 km/h) when diving for prey. They use their talons to grab their prey in mid-air, which they then carry off to eat.

What Bird is That?

Binoculars sorted, the next piece of crucial equipment is a good bird book or field guide. Again, there is an incredible array of field guides: hardbacks, paperbacks, photo guides, artists' illustrations, guides to the birds of Britain, European field guides, specialist books on gulls, waders, rare birds; you name it, there's a book about it. Again, it is largely a matter of personal choice, but if you haven't been birdwatching long, then you may find it easier to begin with a book that focuses just on the birds of Britain and Ireland. This will exclude any confusingly similar-looking species that are found only on mainland Europe, so you can identify with confidence the small, brown bird in your garden as a Dunnock, and not a similar-looking Alpine Accentor that only lives in mainland Europe. Photo guides will show you pictures of the birds as you see them, but field guides with artists' illustrations will accentuate the key identification features you need to look for. The choice is yours. You may find it handy to keep a pocket field guide with you when you're out and about, and a larger field guide with more detailed information at home to refer to when you get back. Have a look at the *Crossley ID Guide: Britain and Ireland*, which is packed with photos, and the *RSPB Guide to British Birds*, which has illustrations. Both come highly recommended.

Take Note!

Keeping a notebook handy can be good practice, to help you get in the habit of not just glancing at birds but really looking at them to study their details. The more you look and notice, the better you will get at bird identification. You don't have to be an artist to do a quick line sketch of what you see, e.g. the shape of the head of a Tufted Duck or perhaps how long the tail is compared to the body of a Long-tailed Tit, but it will help you to really familiarise yourself with the birds you are seeing. Or you might want to make notes about a bird you saw flying by, e.g. medium-sized, black-and-white body & wings, flash of red on back of head, swooping flight across field to tree, to allow you to look it up and identify it as a Great Spotted Woodpecker when you get home. It is so much better to make a note at the time than to rely on your memory hours later.

There's Scope for It

As you progress with your birdwatching, you may find that you'd like to add a telescope to your kit to get even more pleasure out of looking at birds. There's nothing like a 'scope for giving an up-close-and-personal look at birds that may be further away. Take your first telescope look at even a common bird like a drake Teal in his breeding plumage, and we defy you not to say 'wow' as you really appreciate the beautiful colours and detailed patterns of his feathers. Once you have tried it, you'll never go back; you'll find you want to take your telescope with you wherever you go for those extra-special views of the birds. Telescopes offer much greater magnification than binoculars, allowing you to see and identify birds that are much further away, or to really focus on the detail of birds that are closer to you. They usually start at twenty times magnification and can go up to sixty times with a zoom lens, compared to eight to ten times for most binoculars. As with bins, 'scopes come with a variety of objective lens sizes. Again, the larger the lens is, the greater its light-gathering capability, but also the heavier its weight, so you'll want to choose a 'scope that suits your type of birdwatching and fits your budget. As an example, Ruth uses a 25–50 x 65 mm 'scope, while Alan prefers the larger but heavier 25–50 x 82 mm 'scope.

Of course, a telescope needs something to support it. Back in the 1970s, it was considered uncool to use a tripod; birders would rest their telescopes on any available structure to keep it steady, including walls, gateposts, other birders' heads, or would even lie down to rest them on their own feet! Nowadays, life is much easier as people generally use a tripod to support their 'scope. These have adjustable legs so you can set them to the right height for you when looking at birds, and fold them away when you've finished.

Kit and Caboodle

Of course, as with any hobby, once you've got the basics covered there are plenty of gadgets that you can add to your collection to help with your birdwatching. You might want to start photographing the birds you see. Whether you want to grab a point-and-shoot camera or a bridge camera, or invest in a digital SLR camera and a range of lenses, there will be something to suit your photographic aims and budget. Now you can even get specifically designed adaptors with which to attach your smartphone to your telescope, so you can go 'phone-scoping'. It seems everyone is going bird-photography-crazy! If you want bird identification at your fingertips, there are specialist apps for your smartphone to download. And if you want to keep abreast of the latest bird sightings, you can subscribe to bird-information services that send news to a dedicated pager or straight to your phone. Nowadays, it seems the sky, and your budget, is the limit!

CHAPTER 3

GETTING STARTED

Birdwatching is a state of being, not an activity…
it is about life and it is about living.
SIMON BARNES

You've got your binoculars and your field guide ready, so what do you do next to start birdwatching? Well, quite likely you already are birdwatching without realising it.

Birds are everywhere around us, although you may not have paid them much attention before. On a simple walk to the shops you may well pass House Sparrows chirping from within hedges, Starlings whistling from telegraph wires, and Wood Pigeons cooing from rooftops. Walk through your local park and even the most urban pond is likely to have a population of ducks; the majority of these will be Mallards. You might be lucky enough to have a Mute Swan or two there too. Stop in a supermarket car park and you may see a black-and-white bird running around on the ground chasing insects and jerking its tail frenetically up and down: a Pied Wagtail. Or if you're on the coast, a stroll along the promenade may well be accompanied by a Herring Gull looking hopefully to see if you have any spare fish and chips.

Look Where You're Going

The key to starting your birdwatching is simply to look. Start paying attention to your surroundings wherever you are and you will begin to notice birds everywhere. It doesn't matter if you don't know what kind of birds they are; all you need to do is see them and enjoy them. And the more you start noticing your surroundings, the more rewarding it becomes because you will notice more birds. They will probably have been there all the time, but perhaps you haven't registered them before.

So get into the habit of looking where you're going. Or, in fact, don't! Instead of focusing on the path ahead of you and striding out as fast as you can, try slowing down a little. Look all around you – look left and right, and up and down, and you'll start to notice more things – more trees, more bushes, more colour, more movement and, above all, more birds. If you bump into a litter bin in front of you because you were too busy looking right at a Robin singing from a bare branch, that's great! You've just been birdwatching. Stop dead in your tracks to see what's making that bush vibrate so vigorously, and you may be rewarded with the sudden view of a Blackcap as it breaks for cover from one bush to another. How sudden and how delightful! The rest of your walk will be so much better because you had a wonderful and unexpected encounter with a bird. Of course, we don't recommend stepping out into the road with your head screwed round the wrong way, and if you're driving your car, you really should focus on what's in front of you, but other than that, we say just keep looking everywhere, not just where you're going.

Birdwatching can even make doing the housework more fun. If you're lucky enough to have a window over your kitchen sink, then grab your Marigolds and get birdwatching while you're doing the washing-up. You'll be surprised at how many birds you might see in your garden while scrubbing the pots and pans. Again, it doesn't matter if you don't know what they are, just enjoy watching what they're doing as they go about their daily lives. You'll be amazed just how much you will learn about bird behaviour and how much pleasure you'll get while doing so. Without realising it, you'll start to accumulate bird knowledge such as what type of bush attracts what type of bird, what rooftops Collared Doves prefer and what time your Blue Tits start visiting your garden feeder in the morning. All this new knowledge will add to your birdwatching pleasure.

Take Your Bins

Before you know it, you'll be automatically putting your binoculars round your neck whenever you go out for a walk. Having your bins with you really adds to your walk as you'll get much better views of every bird that you see. It is a myth that all birdwatchers have long beards, wear waxed jackets and are nerdy. Most birdwatchers are everyday people who simply enjoy looking at birds, so wear your bins with pride – you'll be surprised how many other people want to talk to you and share their enthusiasm for birds with you simply because you're wearing them. Watching birds while you're out for a walk is very easy: simply stop, lift up your bins and enjoy the bird. It doesn't matter if you don't know what species it is, just smile and appreciate that your walk has been enriched by the encounter.

In the beginning, you'll find it easiest to look at birds sitting out in the open, such as on a bird table or on the lawn in your garden. Carefully note the position of the bird with your naked eye before focusing your binoculars on the bird as quickly as you can. This takes a bit of practice, but the more often you do it, the quicker you'll become. Once you've got your target bird in your bins, simply watch it, taking in the details of its overall shape, colour, bill, wings, feet and any other obvious features. Watch it until it flies away and keep watching until it goes out of sight. Do this every time you see it and you'll gradually build up a complete picture in your mind of what it looks like when it's perched and when it's flying. Then, when you see the same bird in a bush or tree, it doesn't matter if you can't see the whole bird, your memory will fill in the blanks. Watch how it moves amongst the branches or on the ground. Does it hop or does it walk? Does it move about continuously or does it go run-and-stop, run-and-stop? All these different behaviours make watching wild birds endlessly fascinating, no matter how long you've been doing it. People often ask us if we get tired of birdwatching as we've done so much of it. The answer is always no! Every day and every bird is different, and all birds are fascinating to watch.

Commonly Occurring Birds in the United Kingdom

This table gives a list of the birds you can see in the UK. If you're the kind of person who likes to keep track of what you've seen (and if you don't think you are right now, in time you might be surprised!), use the check boxes to mark off the birds as you spot them.

Arctic Skua	☐	Coal Tit	☐	Garden Warbler	☐
Arctic Tern	☐	Collared Dove	☐	Garganey	☐
Avocet	☐	Common Buzzard	☐	Glaucous Gull	☐
Balearic Shearwater	☐	Common Crane	☐	Glossy Ibis	☐
Barn Owl	☐	Common Crossbill	☐	Goldcrest	☐
Barnacle Goose	☐	Common Guillemot	☐	Golden Eagle	☐
Barred Warbler	☐	Common Gull	☐	Golden Oriole	☐
Bar-tailed Godwit	☐	Common Redshank	☐	Golden Pheasant	☐
Bean Goose	☐	Common Redstart	☐	Golden Plover	☐
Bearded Tit	☐	Common Rosefinch	☐	Goldeneye	☐
Bee-eater	☐	Common Sandpiper	☐	Goldfinch	☐
Bewick's Swan	☐	Common Scoter	☐	Goosander	☐
Bittern	☐	Common Snipe	☐	Goshawk	☐
Black Grouse	☐	Common Tern	☐	Grasshopper Warbler	☐
Black Guillemot	☐	Common Whitethroat	☐	Great Black-backed Gull	☐
Black Redstart	☐	Coot	☐	Great Crested Grebe	☐
Black Tern	☐	Cormorant	☐	Great Egret	☐
Blackbird	☐	Corn Bunting	☐	Great Grey Shrike	☐
Blackcap	☐	Corncrake	☐	Great Northern Diver	☐
Black-headed Gull	☐	Cory's Shearwater	☐	Great Shearwater	☐
Black-necked Grebe	☐	Crested Tit	☐	Great Skua	☐
Black-tailed Godwit	☐	Cuckoo	☐	Great Spotted Woodpecker	☐
Black-throated Diver	☐	Curlew	☐	Great Tit	☐
Blue Tit	☐	Curlew Sandpiper	☐	Green Sandpiper	☐
Bluethroat	☐	Dartford Warbler	☐	Green Woodpecker	☐
Brambling	☐	Dipper	☐	Greenfinch	☐
Brent Goose	☐	Dotterel	☐	Greenshank	☐
Bullfinch	☐	Dunlin	☐	Grey Heron	☐
Canada Goose	☐	Dunnock	☐	Grey Partridge	☐
Capercaillie	☐	Egyptian Goose	☐	Grey Phalarope	☐
Carrion Crow	☐	Eider	☐	Grey Plover	☐
Cattle Egret	☐	Ferruginous Duck	☐	Grey Wagtail	☐
Cetti's Warbler	☐	Fieldfare	☐	Greylag Goose	☐
Chaffinch	☐	Firecrest	☐	Hawfinch	☐
Chiffchaff	☐	Fulmar	☐	Hen Harrier	☐
Chough	☐	Gadwall	☐	Herring Gull	☐
Cirl Bunting	☐	Gannet	☐	Hobby	☐

Honey Buzzard	☐	Manx Shearwater	☐	Red-breasted Flycatcher	☐
Hooded Crow	☐	Marsh Harrier	☐	Red-breasted Merganser	☐
Hoopoe	☐	Marsh Tit	☐	Red-crested Pochard	☐
House Martin	☐	Marsh Warbler	☐	Red-legged Partridge	☐
House Sparrow	☐	Meadow Pipit	☐	Red-necked Grebe	☐
Iceland Gull	☐	Mediterranean Gull	☐	Red-necked Phalarope	☐
Icterine Warbler	☐	Melodious Warbler	☐	Redpoll	☐
Jack Snipe	☐	Merlin	☐	Red-throated Diver	☐
Jackdaw	☐	Mistle Thrush	☐	Redwing	☐
Jay	☐	Montagu's Harrier	☐	Reed Bunting	☐
Kentish Plover	☐	Moorhen	☐	Reed Warbler	☐
Kestrel	☐	Mute Swan	☐	Richard's Pipit	☐
Kingfisher	☐	Nightingale	☐	Ring Ouzel	☐
Kittiwake	☐	Nightjar	☐	Ring-billed Gull	☐
Knot	☐	Nuthatch	☐	Ringed Plover	☐
Lapland Bunting	☐	Osprey	☐	Ring-necked Parakeet	☐
Lapwing	☐	Oystercatcher	☐	Robin	☐
Leach's Storm Petrel	☐	Pallas's Leaf Warbler	☐	Rock Dove	☐
Lesser Black-backed Gull	☐	Parrot Crossbill	☐	Rock Pipit	☐
Lesser Spotted Woodpecker	☐	Pectoral Sandpiper	☐	Rook	☐
Lesser Whitethroat	☐	Peregrine Falcon	☐	Roseate Tern	☐
Linnet	☐	Pheasant	☐	Rough-legged Buzzard	☐
Little Auk	☐	Pied Flycatcher	☐	Ruddy Duck	☐
Little Bunting	☐	Pied Wagtail	☐	Ruddy Shelduck	☐
Little Egret	☐	Pink-footed Goose	☐	Ruff	☐
Little Grebe	☐	Pintail	☐	Sand Martin	☐
Little Gull	☐	Pochard	☐	Sanderling	☐
Little Owl	☐	Pomerine Skua	☐	Sandwich Tern	☐
Little Ringed Plover	☐	Ptarmigan	☐	Savi's Warbler	☐
Little Stint	☐	Puffin	☐	Scaup	☐
Little Tern	☐	Purple Heron	☐	Sedge Warbler	☐
Long-eared Owl	☐	Purple Sandpiper	☐	Serin	☐
Long-tailed Duck	☐	Quail	☐	Shag	☐
Long-tailed Skua	☐	Raven	☐	Shelduck	☐
Long-tailed Tit	☐	Razorbill	☐	Shore Lark	☐
Magpie	☐	Red Grouse	☐	Short-eared Owl	☐
Mallard	☐	Red Kite	☐	Shoveler	☐
Mandarin Duck	☐	Red-backed Shrike	☐	Siskin	☐

Skylark	☐	Tawny Owl	☐	White-tailed Eagle	☐
Slavonian Grebe	☐	Teal	☐	Whooper Swan	☐
Smew	☐	Temminck's Stint	☐	Wigeon	☐
Snow Bunting	☐	Tree Pipit	☐	Willow Tit	☐
Song Thrush	☐	Tree Sparrow	☐	Willow Warbler	☐
Sooty Shearwater	☐	Treecreeper	☐	Wood Pigeon	☐
Sparrowhawk	☐	Tufted Duck	☐	Wood Sandpiper	☐
Spoonbill	☐	Turnstone	☐	Wood Warbler	☐
Spotted Crake	☐	Turtle Dove	☐	Woodchat Shrike	☐
Spotted Flycatcher	☐	Twite	☐	Woodcock	☐
Spotted Redshank	☐	Velvet Scoter	☐	Woodlark	☐
Starling	☐	Water Pipit	☐	Wren	☐
Stock Dove	☐	Water Rail	☐	Wryneck	☐
Stone Curlew	☐	Waxwing	☐	Yellow Wagtail	☐
Stonechat	☐	Wheatear	☐	Yellow-browed Warbler	☐
Storm Petrel	☐	Whimbrel	☐	Yellowhammer	☐
Surf Scoter	☐	Whinchat	☐	Yellow-legged Gull	☐
Swallow	☐	White Stork	☐		
Swift	☐	White-fronted Goose	☐		

Thumbing Through

Soon, you may find that your pocket field guide is coming with you on your walks too. Putting a name to a bird somehow makes the encounter more complete. Thumbing through the pages until you find a picture of your particular bird may take you some time initially, but you'll soon get the hang of where birds are in your book, and it's the best way to become familiar with the different birds you see. It's always easier to look up a bird while it's still fresh in your memory rather than try to remember what it looked like when you get home.

Bill Oddie OBE was born in 1941. He wrote many comic sketches and scripts for television and radio and is known to many people of a certain age as one of the three Goodies, along with Tim Brooke-Taylor and Graeme Garden. But he has been a keen birder since he was a young boy and younger television viewers are more likely to know him for his appearances in various BBC natural history programmes, particularly the popular *Springwatch* and *Autumnwatch* series. He was awarded an OBE in 2003 for his services to wildlife conservation. In 2005, he was given the British Naturalists' Association's Peter Scott Memorial Award by the BNA president David Bellamy, in recognition of his contribution to the general public's understanding of natural history and conservation. He has also been awarded the RSPB Medal in recognition of his work towards protecting wild birds and countryside conservation.

There are few better places to study bird behaviour than in your garden or backyard.
BILL ODDIE

www.billoddie.com

Boosting Your Chances

Of course, there are some things you can do to increase your chances of seeing birds. Listening out for their calls or songs will often let you know that a bird is nearby long before you see it, particularly if you're walking through woodland. So walking along quietly will mean that you are more likely to hear and see birds going about their daily business. Wearing subdued colours can be an advantage too. We're not advocating dressing yourself top-to-toe in camouflage, though there are plenty of birdwatchers who do this. Each to their own! But wearing darker colours means that you'll blend in better with your surroundings and, being less obtrusive, you'll cause less alarm to the wildlife around you. There is an American book about birding called *Good Birders Don't Wear White*. We wouldn't necessarily go that far, but dressing in fuchsia pink or dayglo orange, striding along the path and shouting at full voice on your mobile phone, is quite likely to have birds dashing for cover to get out of your way!

Not Reserved for Birds

Visiting your local nature reserve is a great way to see even more birds. These days, nature reserves are increasingly family-friendly and geared to help everyone get the most out of their visit. Joining a guided walk around the reserve is a great way to learn about the place and the birds you can encounter there. You may find information boards and field guides in the hides, and maybe even a helpful volunteer as a 'Guide in the Hide' to point out the birds on view. Hide etiquette is much more relaxed these days. Gone are the days when conversation was frowned upon and hides were as silent as a public library, while opening your field guide to check a bird was considered a real no-no. Nowadays people are keen to share 'scope views and give directions to particular birds or offer identification tips so everyone can enjoy them. Shouting still isn't a great idea, nor is pointing your arm out of the window as you may frighten the birds you've come to see, so simply pull up a bench, lift your bins, open your bird book and enjoy!

Birdwatchers of a Feather, Flock Together

Birdwatching can be a social activity too. Joining your local birdwatching club or society is a great way to meet like-minded people with a shared enthusiasm for watching wild birds. Most societies arrange entertaining talks through the 'indoor season' and organised field trips in spring, summer and autumn, so you can visit new areas, see new birds and make new friends at the same time.

Amazing Bird Fact

Barn Owls can swallow their prey whole, even up to the size of a large rat. Once they've digested the food they will cough up the indigestible bits of bone and fur in a pellet.

CHAPTER 4

HOW TO IDENTIFY BIRDS

The beginning of wisdom is to call things by their right names.
CHINESE PROVERB

Part 1: Bird Anatomy

Just as night follows day, once you've started observing birds it won't be long before you start wanting to identify and name them. Somehow, being able to give a name to a bird makes the whole experience even better. A striking flash of cobalt blue zipping up a river may be exciting to see, but so much more rewarding to be able to say, 'Wow, look! A Kingfisher!' and even though you may have seen only a blue streak, your mind will fill in the missing details of the long black-and-red bill, the orange belly, the dark-blue wings and that amazing cobalt-blue back.

Names are important; after all, that's how we each identify ourselves as individuals. And the best way to recall people's names and identify everyone from one another, is to remember a person's key features. For example, Mary Green is the tall lady with wavy blonde hair and glasses, while Mary Brown is short and dumpy with straight black hair. And while Mike has a grey beard and always wears a green wax jacket, Mick wouldn't be seen dead without his baseball cap, cool shades and faded denim jacket.

The same principle works also with birds: it is really useful to know the key parts of birds and what they're called. It will help you to focus on the particular part of a specific bird that is different from another species. That way you can make a note of it, look it up in a field guide or ask another birdwatcher about it. For example, being able to specify that your brown bird of prey cruising over the marshes had a white rump will mean that you can identify it from your field guide as a female Hen Harrier rather than the plain brown female Marsh Harrier. Or if someone in the hide comments on the striking supercilium on the bird perched on top of a gorse bush, you know to look for an obvious stripe above its eye, and you're halfway to recognising a Whinchat from a Stonechat.

This diagram shows the parts of a bird and what they are called. It's known as bird topography and it allows you to name the different technical parts that are common to all birds. While it's not really necessary to know all the names of the different feathers, knowing what the main body parts are called is useful.

Then, when you see a bird that you don't know, you can start with the bill and work your way along its body to the end of its tail, making a note to yourself either mentally or in your notebook of the key features to look up in your field guide. So that bird with a 'stubby black bill, black crown, pinky-red throat and belly, white vent, grey mantle and back, white rump, black wings with a single white wing-bar and black tail, sitting in an apple tree picking at the buds' can be readily identified as a male Bullfinch, and your pleasure in watching it becomes even greater knowing that it is one.

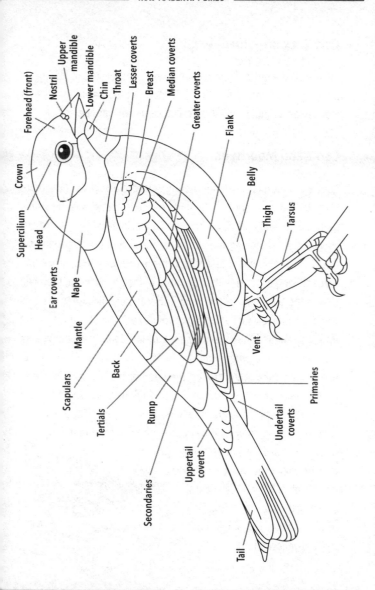

Part 2: Confusion Species

Most birds are quite distinctive once you get the hang of looking at them, but there are a few that you might encounter regularly that look disconcertingly similar: so-called confusion species.

Coots and Moorhens

Coot and Moorhen are both medium-sized black birds that you're likely to find on a pond or small lake, and at first glance they look quite similar. But take a closer look, and you'll realise they are in fact very different characters indeed. Coots are all about swagger: they're the larger of the two, dressed all in black with pale legs, a pale bill and a striking white shield on the front of their head above their bill. They're the bully boys of the pond; they can be quite aggressive and won't hesitate to chase much larger birds such as Canada Geese out into open water if they feel they are too close to their nest or young. Moorhens, on the other hand, are much more secretive and tend to lurk amongst the reeds around the edge of the pond. Even if they do decide to swim right across the middle, they still seem to do it in a quiet, if-I-tiptoe-across-nobody-will-notice-me kind of a way. They're not really black all over either: their head, throat and belly are really a midnight blue, with dark-brown wings, back and tail, a white broken line on their flanks, yellow legs and, most startling of all, a bright red shield and a red-and-yellow bill.

Rooks and Carrion Crows

Two more black birds that are very similar at first glance are Rook and Carrion Crow. These are both members of the corvid, or crow, family that also includes Magpies and Jays, and both birds can be found across most of Britain. Again, they may look very similar as they perch at the top of a tall tree or probe a recently ploughed field for tasty grubs to eat. But, once again, take a closer look and you will start to notice the differences. Rooks have a slightly scruffy and baggy appearance as if their all-black suit of feathers is one size too large for them. See them in bright sunshine and you may notice there is a slight purplish sheen to their plumage. But perhaps their most obvious feature is a pointed, pale, off-white bill. Carrion Crows, on the other hand, are much more dapper chaps, with a well-fitting glossy black suit of feathers and an all-black bill. It's not always the case but it is more common to find one Carrion Crow on its own, whereas a gathering of black corvids is more likely to be of Rooks. As the old country saying goes, 'One rook is crows and two crows is rooks.' Thinking of the noise of a rookery as these birds nest close to each other in the tops of tall trees may also help you remember which is which.

Swallows and House Martins

Identifying birds in flight can be trickier than on the ground, and two confusingly similar aerial masters are Swallows and House Martins. They are both a similar size and shape with blue-black backs and swept-back wings and they both fly in a similar way as they soar and swoop through the sky chasing insects. So far, so similar. However, again, a closer look will show you the differences in their plumages. Swallows have a red face and throat, a blue collar and a white belly, and long tail streamers. House Martins, on the other hand, are completely white underneath with no red face. Their tail is short and forked, but perhaps their most distinctive feature is a white rump which really shows up as they fly around. If you have cup-shaped nests under the eaves of your home, they are likely to belong to House Martins; if you find a nest tucked into the corner of a beam in a barn, it will be a Swallow's nest. Either way, don't worry about any mess; you can count yourself lucky to be blessed with wild birds finding your home as attractive as you do!

People also ask about confusing Swallows and Swifts but these two birds really are different. Granted, Swifts do also soar and swoop through the sky as they hawk for insects. But they are larger and dark brown (often appearing black) all over, with a pale throat and long, sickle-shaped wings. Swifts are the birds you may hear screaming as they swoop low over the rooftops catching insects on warm summer evenings.

Song Thrushes and Mistle Thrushes

Keeping to the theme of confusing birds you may encounter close to home, how about the Song Thrush and the Mistle Thrush? Both are upright medium-sized birds with brownish backs and speckled breasts. The Song Thrush, however, has a warm brown colour on its back, and its breast is yellowy-buff at the top fading to white lower down and speckled with arrow-shaped black spots. You may see it on your lawn or sitting up in a tree singing. Its song is very distinctive as it sings in triplets, i.e. it repeats each particular phrase three times before moving on to the next one. A Mistle Thrush, on the other hand, is a slightly larger bird in much greyer and more faded tones: its back is grey-brown, its underparts are off-white with no yellow, and its black spots are round. Its song is slower and less-often repeated, sounding more like that of a wistful Blackbird, and you are more likely to see it on open fields.

Chiffchaffs and Willow Warblers

Two birds that are so subtly different in appearance that they're easier told apart by their songs are Chiffchaff and Willow Warbler. Both are mostly migrants that appear in Britain in spring to breed, having spent the winter further south in warmer countries, though some hardy Chiffchaffs will overwinter here. Both are small warblers, greenish-brown on the back, paler below and with a pale supercilium (Where's that? Quick, check the diagram!) and, it must be admitted, they're pretty unexciting to look at and very, very similar. Even the experts can get tripped up in separating them if they don't get a good look at them, but they are easily distinguished as soon as they start to sing. A Chiffchaff will very obligingly sing its own name, going 'chiff-*chaff*, chiff-*chaff*, chiff-*chaff*' almost without pausing for breath. A Willow Warbler, on the other hand, has a charming, floaty song which works its way gently down the scale. Yes, it is true that a Willow Warbler has slightly longer wings (which may help with its longer migration deeper into Africa), though this feature can be hard to see clearly. A Chiffchaff will nearly always pump its tail up and down as it flits through the branches, though a Willow Warbler may also do this occasionally so it can't be relied on as a diagnostic ID feature. Just as you spot a tail-pump and decide it's a Chiffchaff, so the bird opens its bill and out floats that gently descending song, declaring it to be a definitive Willow Warbler. So this really is one case where knowing a few key bird songs will help you with your bird ID.

Bar-tailed Godwits and Black-tailed Godwits

You might overhear someone talking about Bar-tailed Godwits and Black-tailed Godwits and take a look at a field guide to find yourself confronted with two more similar-looking birds. How do you tell these two apart? Location can be of some help here, as Bar-tailed Godwits are more generally found in saltwater lagoons and estuaries, while Black-tailed Godwits occur more frequently in freshwater pools, but birds being birds, they don't always stick to the rules, so you might find both species in the same location. Both are biggish waders with long legs, long necks and long bills. Just to make life harder, both species have completely different winter and breeding plumages, and although one does have a solid black band on its tail while the other has black bars on a white tail in all plumages, you need to have an exceptionally good view to distinguish them. This time, it's their overall appearance and movement that will help you tell them apart.

Firstly, Black-tailed Godwits have longer legs. If it were acceptable birdwatcher behaviour, you would have space to write the word 'black' on their legs between their 'knees' (though technically these are their ankles as their real knees are much higher and hidden under their feathers) and their body. Bar-tailed Godwits have shorter legs, with only enough room to write the shorter word 'bar' in the same space. Secondly, Black-tailed Godwits have longer, dead-straight bills. Perhaps because of their longer bills, they feed by reaching forward slightly in front of them to search in the mud for invertebrates to eat. Bar-tailed Godwits, on the other hand, have shorter bills with a very slight up-turn to them. When they feed, also looking for invertebrates to eat, they probe with their bills straight down into the mud between their feet, up and down, up and down rather like miniature sewing machines. You can make out this different feeding technique even at quite a distance, allowing you to say with confidence, 'There's one Bar-tailed Godwit amongst those Black-tails, you know,' before basking in the admiring glances of your suitably impressed friends.

Sparrowhawks and Kestrels

Perhaps you see a bird of prey flying overhead and aren't sure if you're watching a Sparrowhawk or a Kestrel. Again, a good look will help you tell them apart. A Kestrel is a falcon and as such has long, thin, pointed wings. It also has vertical streaks on its breast. You are likely to see it hovering in the air, flapping its wings and making slight adjustments with its tail to hold itself steady and keep its head perfectly still as its gaze is focused downwards, perhaps on a small rodent, its preferred food, scurrying through the grass below. A Sparrowhawk, being a hawk, has broad, rounded wings, and its whole underparts are covered with horizontal bars. It uses the element of surprise to hunt so it tends to sneak low and fast along a hedgerow, suddenly flipping over from one side to the other to ambush an unsuspecting small bird, its preferred prey. And if it's just dashed through your garden, scattering all the Blue Tits from your birdfeeder, then it's definitely a Sparrowhawk!

Chris Packham was born in 1961 and has been passionate about birds and wildlife since a young age. He has appeared in a number of television wildlife programmes, but was perhaps best known for the 1980s children's programme *The Really Wild Show*, until he joined the *Springwatch* panel in 2009. He is known for adding the aspects of scientific background information to *Springwatch* and *Autumnwatch*, and in 2011 he was awarded the Dilys Breese Medal by the British Trust for Ornithology (BTO) for his 'outstanding work in promoting science to new audiences'. He is also known for his sometimes outspoken comments about wildlife issues, such as his views on the Giant Panda captive-breeding programme, and the risks to the planet caused by human overpopulation. He is an active campaigner on environmental issues and holds a presidential position in a number of significant wildlife and conservation charities. He is an award-winning photographer who has applied his own skills as a judge in various photographic competitions.

The speed of the sparrowhawk, the twist of its wings, the brevity of the moment, the feathers floating on the breeze, leave me with an incredible sense of romance.

CHRIS PACKHAM

www.chrispackham.co.uk

Amazing Bird Fact

The Goldcrest is Britain's smallest bird. It measures approximately 3.5 inches (9 cm) long and weighs in at 0.16 ounces (4.5 g). That's about the same weight as a 20p coin.

Part 3: That Fits the Bill

As you get into the habit of looking at more birds, so you'll notice the variation in the shape of their bills and feet. Each bird species has a bill and feet that are designed for the way it lives, and in particular how it feeds. This variation allows lots of different birds to live side by side in the same habitat without competing head-on for the same food. Each bird has developed a unique set of equipment for its own particular niche. Even where you see several similar birds in the same environment, each one will be slightly different to give it a specific advantage. So, for example, if you find several similar-looking wading birds all probing the mud of an estuary, you'll see that those waders with short bills and legs such as Dunlin will be feeding at the water's edge or in the shallows, while those that have longer legs and longer bills such as Black-tailed Godwits will be able to feed further out into deep water. They are all feeding on invertebrates in the same mud, but because of their adaptations, they don't compete directly for the same meal.

You'll notice that there are specific bill designs that are adapted to do a particular job. Here are a few of the most common ones:

Bills

Type: Cracker

Adaptation: Seed-eating birds like House Sparrows and Greenfinches have short, stout conical bills for cracking seeds.

Type: Billhook

Adaptation: Birds of prey like Peregrines, Sparrowhawks and Kestrels, and owls such as Tawny Owls and Barn Owls have sharp, hooked bills for tearing meat.

Type: Chisel

Adaptation: Great Spotted and Green Woodpeckers have bills that are long and shaped like a chisel for boring into wood to find insects to eat.

Type: Probe

Adaptation: Wading birds have long, slender bills for probing into mud for invertebrates lurking within.

Type: Strainer

Adaptation: Some dabbling ducks such as Mallards have long, flat bills that strain small invertebrates from the water.

Type: Spear

Adaptation: Grey Herons and Kingfishers have spear-like bills adapted for stabbing fish in pools and streams.

Type: Pincers

Adaptation: Insect-eating birds like Willow Warblers and Chiffchaffs have thin, pointed bills for delicately plucking insects from vegetation.

Type: Multi-purpose tool

Adaptation: Carrion Crows and Magpies have all-purpose bills that allow them to eat fruit, seeds, insects, fish and even small birds and animals.

Bills need to be strong, and some are quite large (picture a Puffin with its big, multicoloured bill), but they also need to be lightweight to allow flight. Bills are made of keratin, the same substance as our fingernails. The top surface of bills may wear away over time, in which case they will be replenished with more keratin from underneath to keep them the correct length and sharpness, similar to how our fingernails continue to grow throughout our lives. The bill is made up of two parts: the upper mandible, which grows out of the jaw and is fixed; and the lower mandible which is hinged. This movement allows birds to be incredibly flexible and dextrous in using their bill to feed, whether it is picking up tiny individual seeds from a seed head, juggling a fish so it faces the right way to be swallowed or passing small morsels of food to their chicks in the nest.

Not Treading on Each Other's Toes

Different bird species also have differently shaped feet, which are adapted for the particular habitat in which they live, what food they eat and how they eat it.

Here are some examples of the most common feet shapes:

Type: Grasping

Adaptation: Birds like Golden Eagles use their large curved talons to grab their prey and immobilise it.

Type: Scratching

Adaptation: Grey Partridges and other game birds have long, rigid toes for scratching the surface of the soil to find their food.

Type: Swimming

Adaptation: Waterfowl such as Mallards, Mute Swans and Canada Geese all have webbing between their toes to create a paddle for swimming.

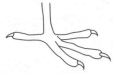

Type: Perching

Adaptation: Perching birds such as Greenfinches and Robins usually have three forward-facing toes and one long back toe for firm gripping so they can perch on tree branches.

Type: Climbing

Adaptation: Tree-climbing birds such as Great Spotted Woodpeckers have strong forward- and backward-facing toes so that they can climb straight up without overbalancing backwards.

Birds' feet are covered in tough scales and are mostly tendons and bones. In fact, what looks like a bird's ankle is in fact only the end of its toes; unlike ours, the joint midway up its leg is in fact its ankle (not a knee bending the wrong way), while its real knee is up under its body, mostly hidden from view. A bird's feet (or rather, its toes!) don't have many nerves or muscles, which means that it does not feel the cold ground or water in the same way as ours would. In fact, the blood vessels in birds' legs are designed to minimise heat loss in this way.

So studying birds' bill shapes and feet adaptations gives you an insight into where they live and how they feed, all adding to your bird knowledge and the sheer pleasure of watching them go about their daily lives.

Part 4: Feathering Your Nest

Many people of a certain age first got into birdwatching through finding nests and collecting eggs as a young child. Even celebrity birdwatcher Bill Oddie admits that as a young lad he started out as an egg collector. Nowadays, egg-collecting is taboo and is even a crime where certain protected species are concerned. Nonetheless, it is interesting to understand why some birds make certain types of nest or lay particular kinds of egg.

Just as not all people like to live in the same type of house, so a bird's nest will reflect its lifestyle and habitat. Some birds, like Blackcaps, make a flimsy structure of dead grasses lined with animal hair. Others are more elaborate, such as the tiny Goldcrest, which makes a nest of moss and spiders' webs slung underneath a branch as far from the trunk as possible. A female Eider duck will make her nest close to the sea and line it with soft downy feathers plucked from her breast to keep the nestlings warm and cosy. Some youngsters have a harsher start in life, though. For example, Ringed Plovers simply make a bare scrape on a shingle beach. Goldeneyes coddle their eggs in a cosy nest hole high in a tree trunk. However, on the same day as the chicks hatch, the mother encourages them to leap out of the nest and freefall down to the ground before making their way on foot to the nearest patch of water and safety. As a Goldeneye chick, you have to hope that your parents have chosen a tree with plenty of soft leaf-litter at its base and only a short walk from water!

Some chicks don't even get a nest in which to start life. Guillemots like to breed in communal seabird colonies, where they congregate in large numbers on the sea cliffs at locations such as RSPB South Stack on the Isle of Anglesey. During the breeding season from April to July, every single narrow ledge on the cliffs is packed with tens of thousands of Guillemots, squashed in cheek by jowl. They lay their eggs directly onto the narrow ledges without any protection at all. As a precaution against rolling off the ledge, the pale green eggs taper to a point at one end. Just like a Weeble, they may wobble, but they won't fall down! Once they hatch, the nestlings also have to balance on the narrow ledges amongst all the other birds while their parents hustle to and fro, bringing in fish for them on a regular basis. After only a few weeks, the adults encourage them to take to the water. The young birds have to hurl themselves off the cliffs to land on the sea maybe as much as 200 feet (61 metres) below. That's quite a leap of faith! They will then stay at sea until they return to land to breed.

Some birds don't bother to make any kind of nest or even involve themselves in rearing their own chicks: the so-called parasitic species. The most famous of these in Britain and Ireland is the Cuckoo. It likes to lay its eggs in the nests of much smaller birds such as Meadow Pipits, Reed Warblers and Dunnocks. Incredibly, the Cuckoo will lay an egg that exactly matches the colour and patterning of its specific host species. A female Cuckoo which parasitises a Meadow Pipit nest with an egg of her own will always target Meadow Pipit nests and lay an egg that looks similar to that of the host. Equally, a Dunnock-specialist will only lay Dunnock lookalike eggs. It's all a case of evolving genetics.

In order to be successful, the female Cuckoo needs to be nippy. In the space of just a few seconds, she will land on an unattended nest, eject one of the original eggs, lay her own and then depart the scene of the crime before the unsuspecting host bird returns. The host parents, let's assume Meadow Pipits in this case, will incubate all the eggs without noticing that anything has changed. The Cuckoo egg will usually hatch before the others and the chick's immediate instinct is to lever the other eggs (and any Meadow Pipit chicks that may have hatched first) over the side of the nest so that it is the only remaining chick in the nest to be fed. The Meadow Pipit parents now have a full-time job feeding a chick that quickly grows to become much larger than they are, although they don't seem to notice this. In fact, the Cuckoo chick even makes a begging call that sounds like a nest-full of hungry young, just to keep the parents returning with more food. The Cuckoo chick will finally fledge after about 21 days, and will soon become self-sufficient. The poor exhausted Meadow Pipit parents presumably lie down in a darkened room somewhere to recover!

Amazing Bird Fact

About half of a bird's head is taken up by its eye. Our eyes only take up about 5 per cent of our head space, so if our eyes were the same as birds', they would be as big as tennis balls.

Part 5: Name That Tune in One!

Certain things we associate with summer: warm weather (maybe), sunshine (if we're lucky) and birds singing (best of all); but while birdsong might add to our own feeling that all's well with the world, that's not really why birds are singing. For them, it's all about communication.

Birds use their voices to communicate with one another and, just like human speech, different types of sound send different kinds of message. Bird calls can be heard all year round. These may be short contact sounds such as *tick* or *chack* and, while many of them may sound similar to our ears, each call will be unique to each species and is conveying a particular message. For example, they may be communicating directly from one bird to another, such as between a parent bird and its fledgling so they both know where the other is, even if they can't see each other in the vegetation. Or perhaps it may be to broadcast generally to warn of a predator threat. Other birds may hear the alert and know to be on the lookout, either by taking cover or by taking action to oust the threat. For example, if a Blackbird spots a Tawny Owl roosting in a tree during daytime, it will set up such a loud commotion that it attracts the attention of other smaller birds who join in the mobbing, and they collectively harry the owl until it is driven away to find somewhere quieter to sleep.

Birdsong is a different proposition altogether. It's usually the males who sing, to proclaim their territory and deter rivals (this is mine, so keep out!) and/or to attract a female (hello, I'm here, I'm eligible and I've got a really nice patch to offer you!). The work of finding and keeping both a territory and a mate takes place in spring, starting around February and continuing usually until June, which is why you will hear the most birdsong during these months. After this, it all goes quiet on the birdsong front. Once the fledglings (young birds that have left the nest) are old enough to fend for themselves, the adult birds will shed their worn plumage, which is replaced with strong new feathers for the autumn and winter. Losing their feathers makes it harder for birds to escape from predators, so the last thing they will want to do is sing to attract attention to themselves. Of course there are exceptions to every rule, and Robins, both males and females, will continue to sing throughout the winter to protect their territory all year round, and coincidentally to gladden the heart of any birdwatcher on a winter walk.

So there's more to a bird's song than initially meets the ear. First of all, each song is unique to each species, so a male Willow Warbler's song will only attract a female Willow Warbler. It will also give an indication of how fit and well the male is; a long and loud song suggests the male is hale and hearty. In some cases, the relative complexity of a bird's repertoire is also an indicator of its physical condition, so a long, loud and complicated song may suggest that the songster is a real catch for the ladies. It also acts as a deterrent to other would-be suitors in the area, who may not be able to face such stiff competition for a partner.

And why the dawn chorus? Well, there are quite a few theories about why birds sing so much at first light and then fade away as the day goes on. It could be that it proves that the male has survived the night and is still there to defend his territory. Also, the female may be at her most fertile at dawn, having just laid her eggs. It may be that there is less noise pollution from other sources then, or it could be the best use of time until it is light enough to find food. Whatever the scientific reason, hearing the dawn chorus of birdsong is definitely the just reward for any birdwatcher that has got out of bed at first light to look for birds.

From a birdwatcher's point of view, birdsong is a really helpful clue as to what birds are about. The majority of singing birds are passerines, or perching birds. That means they are most likely to be found in areas where there are trees, bushes and other cover on which to perch. Once the leaves are out in spring, it can be much trickier to see birds up in the branches, so hearing them sing will let you know they are there even when you can't see them. In fact, you are much more likely to hear them singing first, and once you know they are there you can start to look for them. So recognising even a few key bird songs will help you add to the number of birds you see on your walk.

Some bird calls are so distinctive that everybody knows them, such as that of the Cuckoo. Some other birds also helpfully say their own name, including the Chiffchaff. Others have found their way into popular culture, such as the Nightingale's throbbing song (good luck in trying to hear that over the central London traffic of Berkeley

Square!) or the Turtle Dove's gentle purring, but you would be very lucky indeed to hear either of these. If you can pick out your Robin from your Blackbird, and distinguish between your Wood Pigeon and Collared Dove, Song Thrush and Mistle Thrush, Willow Warbler and Chiffchaff, Blackcap and Garden Warbler, Nuthatch, Coal Tit, Blue Tit and even just some of the many noises a Great Tit can produce, then you're well on the way.

Some birds have very distinctive calls or songs which can be written as follows:

Bittern (male): *boom*
Red Grouse: go-*back*, go-*back*
Collared Dove: uni-ted
Wood Pigeon: a *good*, wood pigeon
Tawny Owl: ke-*wick*
Bearded Tit: *ping, ping*
Great Tit: tea*cher*, tea*cher*
Rook: *caw*
Jackdaw: *jack, jack*
Raven: *kronk*
Chaffinch: *pink, pink*
Crossbill: *chup, chup; chup, chup*
Yellowhammer: a little bit of bread and no *cheeeese*

Part 6: What's in a Name?

Odd Bird Names

Some birds have names that are onomatopoeic, i.e. they sound like the calls that the birds make. Birds that say their own name:

Chiffchaff: chiff *chaff*, chiff *chaff*

Chough: chee-*owgh*

Cuckoo: cuc-*koo*

Curlew: curr-*lew*

Kittiwake: kitty-*waik*

Peewit (old name for Lapwing): *peee*-wit

Bird Names that Fit the Part!

Wryneck: a member of the woodpecker family, belonging to the genus *Jynx*. It has cryptic plumage with an intricate pattern of greys and browns, making it very hard to see. It gets its name from being able to twist its head through almost 180 degrees. It twists its head in this way and hisses as a threat display when it is disturbed at the nest. Wrynecks were apparently used in witchcraft on account of this behaviour, which led to the phrase 'putting a jinx' on someone.

Crossbill: a large member of the finch family. The upper and lower mandible in its chunky bill are crossed over at the tips, which allows it to prise out the seeds it eats from a pine cone.

Snowy Owl: so called because of its snow-white plumage, making it perfectly adapted to live above the Arctic Circle.

Razorbill: a black-and-white seabird, so called because of its thick, flat bill shaped like an old-fashioned cut-throat razor.

Pintail: a very elegant duck; the male has particularly long central tail feathers, which give it its name.

Birds Named After People

If you discovered a new species, common practice was to name it not after yourself but after a close friend or colleague you admired, presumably in the hope or expectation that they would repay the favour!

Bewick's Swan: named after Thomas Bewick (1753–1828), famous for his wood engravings in his book *A History of British Birds*.

Temminck's Stint: named after Coenraad Jacob Temminck (1778–1858), a Dutch aristocrat who was the director of the National Natural History Museum at Leiden. In 1815, he wrote an important book on European birds called *Manuel d'ornithologie, ou Tableau systématique des oiseaux qui se trouvent en Europe*.

Montagu's Harrier: named after George Montagu (1753–1815), who in 1802 wrote the *Ornithological Dictionary* cataloguing the status of Britain's birds.

Cetti's Warbler: named after the Italian zoologist, Francesco Cetti (1726–1778). He wrote the *Storia Naturale di Sardegna* (Natural History of Sardinia), a four-volume work covering mammals, birds, fish, insects and fossils that took him four years to write.

Wilson's Phalarope: named after Alexander Wilson (1766–1813), an American ornithologist, naturalist and illustrator. He is considered the greatest American ornithologist before John James Audubon.

Baird's Sandpiper: named after Spencer Fullerton Baird (1823–1887), an American naturalist, ornithologist, ichthyologist (fish specialist), herpetologist (amphibian expert) and museum curator. He published over a thousand books and papers during his life and built up the natural history collection at the Smithsonian from 6,000 to over two million exhibits.

Leach's Petrel: named after William Elford Leach (1790–1836), a British zoologist and marine biologist who sorted out many of the unattended collections of the British Museum and became a crustacean and mollusc expert, publishing a number of books and papers on the subject. The Petrel was named after him by Coenraad Jacob Temminck.

Knot: not named after a Mr Knot, but its Latin name is *Calidris canutus*, a reference to King Canute and the story of him holding back the tide (or not), and the bird's habit of foraging along the tideline for food.

Collective Nouns for Birds

Birds in groups are described as being in flocks, but some bird species have particular collective nouns for when they gather together. Here are a select few:

- A flight of cormorants
- A covert of coots
- A murder of crows
- A raft of ducks
- A congregation of eagles
- A charm of goldfinches
- A covey of grouse
- A skein of geese (in flight)
- A gaggle of geese (on the ground)
- A colony of gulls
- A siege of herons
- A scold of jays
- An exaltation of larks
- A wisdom of owls
- A colony of penguins
- A nye of pheasants
- A congregation of plovers
- A congress of ravens
- A parliament of rooks
- A host of sparrows
- A murmuration of starlings (in flight)
- A flight of swallows
- A lamentation of swans
- A descent of woodpeckers
- A herd of wrens

Part 7: Doing the Nightshift

Some birds are nocturnal, i.e. they are more active during the night and spend the daylight hours sleeping. Most people are familiar with owls and we have five resident species of owl in Britain and Ireland: Barn Owl, Little Owl, Tawny Owl, Long-eared Owl and Short-eared Owl; but these wonderful birds can be tricky to see. Most owls are nocturnal and hunt at night, though some search for their prey at dawn and dusk and so are described as crepuscular. However, Little Owls and Short-eared Owls can be seen hunting during the day, while during the breeding season, demand for food from hungry chicks is so high that adult Barn Owls can also be seen hunting during broad daylight. All owls are carnivorous and are adapted to live and hunt at night, with highly developed senses of hearing and smell, and specially adapted eyesight. They have large eyes positioned on the front of their heads to make the most of low light levels, surrounded by large facial disks which are in fact feathers adapted to capture more sound, giving owls their particularly acute hearing. Owls' feathers are also adapted to allow them to fly slowly and almost silently, allowing them to sneak up on their prey soundlessly in their night-time world.

Nightjars are also nocturnal birds found on mainland Britain. They breed in undisturbed areas of open heathland where, with luck, you may see them flying low across the heather in warm, calm weather at dusk and dawn. They fly through the air with their bills open as they hawk for insects, their preferred prey. Their old country name was 'Goatsucker', as they were alleged to steal milk from unwary goats and cows, though they are more likely to have been snatching the insects disturbed by the animals' feet than grabbing an illicit milky drink. They spend the day sleeping in full view on open ground, relying on their cryptic, grey-brown, patterned plumage to conceal them. This camouflage is incredibly effective; unless they break cover, you can walk within a few feet of a roosting Nightjar without seeing it. Much like owls, they can fly almost silently, and often the first you will know of a Nightjar in the vicinity will be if the male starts making its bizarre, almost mechanical song known as 'churring'. These rather falcon-like birds arrive back here in May, returning to Africa in September for the winter.

Of course, not all birds that are out and about at night are, strictly speaking, nocturnal. For example, Robins are able to forage in low light, and artificial streetlights may sometimes trick them into singing during the night-time. Other birds, such as migrating waders and passerines, may take advantage of the cover of darkness to continue their journey north or south with less likelihood of being attacked by predators such as Peregrine Falcons, Merlins, Goshawks and Sparrowhawks, all of which specialise in hunting smaller birds. These predators lie in wait along migration routes to pick off tired and weary migrants. Recent research has shown that Peregrine Falcons are evolving to hunt using the artificial light created by our city centres. These amazing birds of prey are able to intercept migrant birds passing over our conurbations by capitalising on the fact that the manmade light sources below are illuminating their prey.

Some seabirds, such as Manx Shearwaters and European Storm Petrels, use the hours of darkness to come ashore to visit their nests, which allows them to avoid predators. Many of these seabirds nest on offshore islands where they are safe from ground predators, but they still run the risk of being attacked by avian hunters such as large gulls, Peregrine Falcons and Short-eared Owls. Many of them are very vulnerable on land; while they are supremely adapted for a life out at sea, they are often awkward, slow and clumsy on the few occasions when they come into land, making them easy targets for predators to pick off.

CHAPTER 5

WHERE TO GO BIRDWATCHING

In all things of nature there is something of the marvellous.

ARISTOTLE

The first, and easiest, way to start your birdwatching is not to go anywhere at all. Simply stay at home and look out of your window. It doesn't matter where you live – whether you're in the heart of the city or deep in the countryside – if you look out your window, you will almost certainly see birds.

We live in a second-floor apartment in a town, so we don't have a garden ourselves, but we still have a long list of birds we've seen from our living-room window. Most of them are flying over the rooftops and include our regulars such as Herring Gull, Jackdaw, Collared Dove, Feral Pigeon, House Sparrow and Goldfinch. Some are much more interesting and our living-room bird list includes Peregrine Falcon, Osprey, Fulmar, Chough, Curlew, a flock of six White Storks and even a real rarity called a Royal Tern on one occasion.

If we look down, we can enjoy the birds that visit other people's gardens, and if you are lucky enough to have a garden you will be able to see plenty of birds without even going out of doors. Plenty of bird species thrive in the kind of 'edge of woodland' type of habitat we have artificially created in our gardens and they have learned to live alongside human beings. Simply looking out of the window while sitting in your armchair, you can watch the birds living out their daily lives in your garden and get to know the species you see on a regular basis.

David Lindo, also known as 'The Urban Birder', was born in 1963 in east London, where he grew up. Despite his potentially uninspiring urban surroundings, David became hooked on birds and ever since then he has been inspired to watch them and to encourage others in an urban area to do the same. 'Just look up' is his mantra as he takes delight in showing people through his radio and TV appearances, articles and talks just how much wildlife is all around them, no matter how built-up the area may be.

The sky is my canvas and it's amazing what you can see when you look up… It's a beautiful world up there.
DAVID LINDO

www.theurbanbirder.com
#LookUp

Gardening for Birds

Feeders

Of course, you can stack the odds in your favour when it comes to encouraging even more birds into your garden. Putting out food for them is the first easy step. Providing a mix of food will encourage the widest variety of birds; after all, we all prefer a restaurant with an extensive menu! These days, there is a staggering array of different feed and feeders to attract birds into your garden. Take a look at the bird-food page on the RSPB website and you'll get the idea: http://shopping. rspb.org.uk/bird-food.html. Your only real limit is your budget, but even if you only provide the basics of a peanut feeder, a seed feeder and some water, you may find your garden becomes the fast-food outlet of choice for all manner of birds, possibly including Robins, Blackbirds, Blue Tits, Great Tits, Coal Tits, Woodpeckers, Nuthatches, Chaffinches, Greenfinches, Goldfinches and more. Sunflower hearts are a fantastic nutritious food that leaves no mess and appeals to the highest number of birds, but they are also the most expensive. However, we think it's a small price to pay for the pleasure of seeing so many birds so close up.

Fat Balls

Birds love fat balls, which can be an important source of energy in winter. You can buy them from a reputable supplier or, if you prefer, you can make them yourself from kitchen scraps such as cheese and dry porridge oats. Just mix with melted lard or suet and set in the fridge overnight. Avoid using turkey fat, which won't set; it can also coat birds' feathers, preventing them from flying. Here's a recipe for wild-bird cake which will have them flocking to your garden!

Ingredients and equipment:
Lard or suet
Bowl
Saucepan
Spoon
String or twine
Old yogurt pots
Any of the following: wild bird seed, currants, sultanas, oats, breadcrumbs, cake crumbs, grated cheese, peanuts

Method:
For the best fat balls you will need one part fat to two parts dry mixture.

Mix all the dry ingredients together in a bowl.

Melt the lard or suet in a pan and then add the dry mix. Stir well until the fat has all been absorbed and the mixture sticks together.

Make a hole in the bottom of a yogurt pot and thread through a length of twine or string, then fill the pot tightly with the warm fat mixture.

Place the filled yogurt pot in the fridge overnight to set, then cut through and peel away the plastic pot and dispose of it carefully. Tie a big knot at one end of the twine to secure the fat ball.

Hang the fat ball in a tree or shrub, ideally not too far from some vegetation to provide cover from watchful Sparrowhawks. Sit back with your binoculars and wait for the birds to come to your food.

It's critical to feed the birds throughout the winter when there is less natural food available for them, and the extra nutrition they get from visiting your feeders could literally mean the difference between life and death on a cold winter's night. In early spring, your bird food will help resident birds and returning migrants put on weight and stay fit and healthy at the start of the breeding season. You may find fewer birds visiting your feeders in summer when there's plenty of natural food available, but before you know it, the temperatures will be dropping at the start of autumn and birds will be queuing up at your feeders once more.

Plants, Trees and Shrubs

You might like to consider adding particular plants to your garden to provide cover and/or another source of food. There are plenty of books on the subject, such as the BTO (British Trust for Ornithology) book entitled *Gardening for Birdwatchers* by Mike Toms and Ian & Barley Wilson, which covers how to design your garden with wildlife in mind. For example, you might want to leave corners of your garden relatively untended – the perfect excuse to be a lazy gardener! They also discuss which particular plants and shrubs will encourage which species, e.g. growing berry-bearing bushes such as pyracantha or rowan trees to attract Blackbirds, Song and Mistle Thrushes, Redwings and Fieldfares, and possibly an exotic Waxwing or two, and how to minimise the risk from predators seeking to capitalise on your new bird-jammed garden by providing plenty of bird-friendly cover.

Nest Boxes

Suitable nesting sites close to food sources can be at a premium for some birds, so by putting up nest boxes you can encourage more birds to raise their young in your garden too. Again, there is plenty of sound advice in books and online about how to go about putting up nest boxes in your garden: for example, www.bto.org/about-birds/nnbw/putting-up-a-nest-box has information on which nest boxes are designed for which species and where best to locate them in your garden. The 'traditional' individual nest box with a round hole will make a perfect home for a Blue Tit to raise its family. House Sparrows may take to a communal nest box with space for several nests side by side. And if you have a really large garden with a stout tree, how about putting up a specialist nest box large enough for a Tawny Owl? You can even get nest cams – specially designed cameras which fit inside your nest box and can beam a live feed back to your computer screen or television, so you can create your very own *Springwatch* viewing and watch the bird-family drama unfold minute by minute from your sofa!

These are the UK's top 20 garden birds according to the results of the RSPB Big Garden Birdwatch in 2014:

Species	Mean	Percentage of gardens	Percentage change since 1979
House Sparrow	3.788	55.19	−62.1
Blue Tit	2.456	73.52	+0.7
Starling	2.375	33.09	−84.2
Blackbird	2.168	82.06	−45.8
Wood Pigeon	1.685	63.21	+742.5
Chaffinch	1.497	41.3	−50.1
Goldfinch	1.426	28.48	NA
Great Tit	1.251	51.43	+39.0
Collared Dove	1.162	44.71	+315.1
Robin	1.099	73.05	−45.1
Magpie	0.918	45.15	+129.6
Dunnock	0.813	43.5	+1.6
Long-tailed Tit	0.768	18.66	NA
Feral Pigeon	0.62	16.3	NA
Greenfinch	0.556	19.67	NA
Jackdaw	0.55	15.24	−44.4
Coal Tit	0.535	28.05	NA
Carrion Crow	0.489	19.62	+181.8
Wren	0.178	15.69	NA
Great Spotted Woodpecker	0.121	9.98	+11.4

Note: NA means that no data was available in 1979

It's interesting to note the winners and losers since 1979, when Big Garden Birdwatch began, with Wood Pigeon making the largest increase and Starling the steepest decline in numbers in our gardens. Do your experiences in your own garden bear this out?

Spreading Your Wings

Parks and Green Spaces

Once you're more familiar with the birds in your garden, you may want to start looking further afield to see what bird species you can discover. You may be lucky enough to have a park nearby. If this has some form of water in it, such as a pond or stream, then you will have a whole new range of species to look for: waterbirds. Many of us have fed the ducks in the park as children; most of these will have been

Mallards, but in amongst the ducks on larger areas of water, you may also find other species such as Tufted Duck, Wigeon, Teal or Pochard in the mix. Moorhens and Coots could be pottering about at the edge of the water, while Reed Warblers and Sedge Warblers may be heard singing in reed beds. Open areas of grass may support Green Woodpeckers probing for ants, while Jays may be hiding acorns for their winter store and Rooks may be holding their parliamentary gatherings in the tops of tall trees. Areas of shrubbery may provide cover for skulkers such as Blackcaps, or birds like Bullfinches that are quite secretive despite their gloriously vivid colouring. Soon you'll find you are taking your binoculars with you every time you go to the park.

Nature Reserves

The next step on your birdwatching journey should be to pay a visit to your nearest nature reserve. Look online and you'll find there are plenty of these across the country. Some are designated Local Nature Reserves managed by the council, others may be part of a network of protected areas managed by charities and NGOs (Non-Governmental Organisations), such as the RSPB and The Wildlife Trusts. All are great places to meet like-minded people who will be happy to help with your bird identification, and of course they are excellent places to enjoy seeing a whole range of birds in a variety of different habitats.

Amazing Bird Fact

The White-tailed Eagle is the largest bird in Britain. Sometimes described as a 'flying barn door', it has an 8-foot (2.4-m) wingspan.

Ten Top Places to See Birds in Britain and Ireland

1. RSPB Titchwell Marsh, Norfolk

Situated on the north Norfolk coast, east of Hunstanton, this wetland reserve always has a lot of birds to see all year round. The trees surrounding the car park hold warblers and thrushes, the feeders by the visitor centre attract tits and finches. Walk out towards the marshes and you might well see a Barn Owl or a Water Rail. The pools, lagoons and reed beds are home to masses of birds, including waders, wildfowl, Bearded Tits, Marsh Harriers and Bitterns. Continue to the beach and you can add seabirds, including gulls, terns and seaducks, to your impressive list. Perhaps the best time to visit is in the autumn when the lagoons can be teeming with migrant waders. This is a sight not to be missed.

2. RSPB South Stack, Anglesey, North Wales

The spectacular scenery here alone makes the visit worthwhile: towering sea cliffs, an island with a lighthouse, coastal heathland and views back across to the mountains of Snowdonia – wow! But the birds are also worth the trip too. This is perhaps the best place to see the red-billed Chough, a rare member of the crow family that can be seen here all year round. If you visit between April and June you will be in for a real treat, as the cliffs will be transformed into a noisy seabird city! Thousands of Common Guillemots, along with Razorbills, Fulmars and small numbers of Puffins make their nests here. Peregrine Falcons cruise over the rock faces and migrants can, and do, pop up anywhere; a long list of rare birds have been seen here.

3. The Isles of Scilly, off Cornwall

Lying off the coast of Cornwall, south-west of Land's End, the Isles of Scilly have long held a magnetic draw for the keenest birdwatchers intent on seeing fabulous rarities. The islands' far-flung location make them a magnet for lost migrant birds too; many American species have ended up here after being blown out over the Atlantic by hurricanes off the east coast of America. But birds from all points of the compass end up on the islands and it really is a case of expect the unexpected. September and October are the prime months for rare birds but the spring months can also produce amazing varieties. More recently, many birdwatchers have started to visit Scilly in late summer to look for rare seabirds during boat trips from the islands. These 'pelagic' trips have produced numerous sights of the very rare Wilson's Petrel – a seabird seldom encountered in our waters.

4. Farne Islands, Northumberland

Just off the Northumberland coast from Seahouses these National Trust-owned islands are home to huge numbers of seabirds from April through to July. Take a boat trip out to the rocky islands and you will be amazed by the spectacle. You can literally walk among the birds, so there will be Puffins at your feet and very likely an Arctic Tern on your head – top tip: do wear a hat! If you like taking photographs of birds then this really is a place you must visit, and take plenty of memory cards. Common Guillemots mass on flat-topped stacks, below them Kittiwakes somehow manage to secure their nests to the rock walls, and Razorbills and Shags occupy any tiny ledge that's available. The sound and smell make for an unforgettable experience. Sandwich, Common and Arctic Terns all breed here and the Arctic Terns will dive-bomb you, even pecking your head in defence of their nests right by, sometimes even on, the paths. Do remember that hat!

5. WWT Martin Mere, Lancashire

This wetland reserve on the low-lying ground east of Southport is an important wintering ground for wildfowl, particularly Whooper Swans and Pink-footed Geese. The swans and geese breed in Iceland and fly south to escape the Arctic winter, making the reserve their temporary home. A visit to the reserve between November and March will be rewarded with the sight of around a thousand Whooper Swans and masses of Pink-footed Geese; the sound is just as impressive as the sight! In the late afternoon, the wild swans are fed in front of the hides allowing wonderful close-up views and photo opportunities. Winter also brings birds of prey. Peregrines are a daily sight, often showing very well, while Merlins and Hen Harriers are also seen. In the summer months, Avocets breed on the lagoons and there is often a Marsh Harrier about.

6. Grantown-on-Spey, Scottish Highlands

If you use this lovely town as a base to explore the Spey Valley and nearby Cairngorm Mountains, you have a lot of amazing birds within easy reach. The ancient Caledonian pine forests here are home to Crested Tits and the sadly rapidly declining Capercaillie. RSPB Loch Garten, with its famous breeding Ospreys, is just a short drive away and a great place to see red squirrels and perhaps even one of those elusive Capercaillies? Take the mountain railway to the top of Cairn Gorm itself and you may well see Ptarmigan from the viewing area, and look out for Ring Ouzels behind the bottom station, also a great place to see Red Grouse. To the north lies Findhorn Valley, where Golden Eagles soar above the ridges. Some of the lochs have breeding Slavonian Grebes and Black-throated Divers and even a few pairs of Goldeneyes. Perhaps May is the best time of year to visit, but the area is great all year round.

7. Dungeness, Kent

Situated on the south coast jutting out into the English Channel, the Dungeness peninsula is a magnet for birds and birdwatchers. You also get two for the price of one here, as Dungeness Bird Observatory is right next to RSPB Dungeness Nature Reserve! Dungeness is ideally placed as a stopping-off point for migrant birds, being one of the first places they can make landfall as they travel north across the Channel in spring. Then, in the autumn, it is one of the last places they can stop to refuel before making the sea crossing south. Many scarce and rare birds are recorded at the observatory and it is an exciting place to witness bird migration in spring and autumn. The adjacent RSPB reserve has lagoons, pools and reed beds, giving another exciting mix of birds to enjoy. There are birds to enjoy here all year round. Again, these wetland habitats have attracted many migrant, scarce and rare birds over the years, so any visit will be filled with great anticipation. In winter it is one of the most reliable places to see Smew, a scarce visitor from Scandinavia: look for these small diving ducks on the lake that is known as the ARC Pit on the RSPB reserve.

8. Rutland Water, Leicestershire

The Anglian Water Birdwatching Centre with its superb facilities is located on the Egleton area of the Rutland Water Nature Reserve at the west end of the lake. It is open daily throughout the year and there are always lots of birds to see. Rutland Water is a huge man-made reservoir and sympathetic management of the lake has made it a haven for birds. Ospreys breed here and many visitors are drawn during the summer to see these amazing fish-eating birds of prey. But there is much more! The open water attracts many diving ducks and grebes, with Black Terns calling in every spring and autumn. The shallow bays and specially constructed lagoons are a haven for dabbling ducks and waders. The extensive network of rails and many hides make this site easy to visit and to enjoy. Rutland Water is also the home of the annual Birdfair (see over).

Tim Appleton MBE is the Reserve Manager at Rutland Water Nature Reserve, where he has worked since its creation in 1975. **Martin Davies** has worked for the RSPB for more than 35 years and was for many years Head of International Funding. However, Tim and Martin are perhaps best known as the dynamic duo who co-founded and continue to organise the annual British Birdwatching Fair, also known as Birdfair. This 'birders' Glastonbury' takes place over three days in August each year at Rutland Water and now attracts over twenty-six thousand visitors, a far cry from its humble beginnings as 'The Wildfowl Bonanza' in 1987. With an ever-increasing number of marquees housing displays of everything and anything to do with birds and birdwatching in Britain and around the world, this is a must-visit event for any birdwatcher. But more than just being a trade fair, Birdfair is a significant fundraising event with monies being raised in aid of a different bird conservation issue each year. Between 1989 and 2013 the event had raised a grand total of £3,396,152.

9. Unst, Shetland Isles

The most northerly point of the British Isles, this far-flung island is a paradise for birdwatchers, given some reasonable luck with the weather! In spring and summer the island is home to tens of thousands of breeding seabirds, many of which make their homes on the towering cliffs. At Hermaness, a huge colony of Gannets clings to the massive cliffs and peering down at them is quite terrifying! To reach the Gannets you cross a moorland where the Bonxie, or Great Skua, holds sway. These huge birds repeatedly dive-bomb you, sometimes making contact (which can be a terrifying experience) as you cross their territory. Puffins line up along the cliff tops, always bringing a smile to everyone's face. In both spring and autumn, Unst is a magnet for lost migrants and vagrant birds; the list of rarities recorded here is long and mouth-watering. To see amazing birds in such a wild and rugged landscape is a dream come true. Though difficult to reach, Unst is a magical island and everyone who makes the long journey falls under its spell.

10. Tacumshane, County Wexford

Tacumshane is a small village in the south-east corner of County Wexford, Republic of Ireland, but it is the nearby shallow lake and marshes that share the village's name that is familiar to birdwatchers. Tacumshane is quite simply a magnet for birds! The shallow, muddy waters here produce rich feeding for wildfowl and in particular wading birds. Looking at the map, it does not seem that the area is particularly well placed to receive migrant birds but somehow they come and stay to feed. Few sites can rival the number of rare and scarce birds that have occurred here over the years. Not just single rare birds, as in most locations, but sometimes quite literally flocks of rare birds! These have included flocks of Buff-breasted Sandpipers all the way from North America, surely having flown over many other Irish marshes before landing here at Tacumshane? The list of American waders and wildfowl recorded here is second to none, and all these rare birds attract birdwatchers to the site. These very birdwatchers then go on to find more unusual birds during their visits and this brings yet more birdwatchers! Tacumshane benefits from having very astute local birdwatchers that comb the area working hard to pick out the unusual from the hordes of commoner birds. September is THE month to visit, especially for American waders, but good birds can and do occur here throughout the year.

11. WWT London Wetlands Centre, Barnes

Yes, we know that we have eleven places in our Top Ten list, but we just had to include this wonderful urban oasis! You might think it unlikely to find a wetland haven in the centre of London, but you'd be very wrong. On the site of a disused Victorian reservoir system just south of the Thames, this WWT reserve at Barnes has offered a calm and tranquil retreat packed with wildlife since its opening in 2000. This is a really family-friendly reserve; paths lead you through a series of landscaped areas with representative birds from different parts of the world. Beyond this, you reach the wild landscaped wetlands, with hides offering you great views of the birds, including some species that you will not find elsewhere in London. There are nationally significant numbers of Gadwall and Shoveler here, as well as Little Grebes, Great Crested Grebes and Lapwings. If you're really lucky, you may also catch sight of the elusive Bittern and Water Rail here, while an azure-bright Kingfisher may put on a show for you. And after all that urban birding, there's a terrific cafe to sustain the inner birdwatcher!

Top Ten Global Birdwatching Destinations

Luckily birds are found, and can be enjoyed, all around the world, so picking a list of top ten destinations is a very personal thing. Ask any birdwatcher for their top ten and we are sure you would have as many answers as there are birdwatchers!

So our list is just that, ten countries in which we love birdwatching. But we have applied some 'rules' to our selection. These rules mean that each country chosen ticks the following boxes: great birds, of course; a good, safe infrastructure when it comes to hotels and eating; and a good field guide so you can identify the birds. So those are the ground rules and here are our top ten places to go birdwatching globally.

Finland and Norway

⑨ Thailand

amibia

⑩ Australia

1. Britain and Ireland

Well, we had to include our home country! We have a wealth of habitats with lots of birds in a small area. We lie at a bird crossroads, with arrivals from all points of the compass making the migration season particularly exciting and unpredictable. A network of nature reserves and a wealth of information make birdwatching very enjoyable and accessible to all.

Our personal bird highlight: it's got to be Puffin. We're lucky enough to have these wonderful, clown-faced birds breeding near us in summer, and they bring a smile to our faces every time we see one.

2. Spain

We have been lucky enough to visit this amazing country many times and love the birds here. The species are different from our own, yet familiar to us through our European Field Guides, so not an overwhelming experience. The scenery is varied and often dramatic, with a wealth of birds of prey, which are always a thrill to see. Easy to reach and relatively inexpensive, it is a perfect first overseas destination.

Our personal bird highlight: Great Bustard. The male bird can transform himself from a brown turkey-sized bird into a huge white puffball by puffing up his throat and turning his feathers almost inside out to attract the females!

3. Finland and Norway

Yes, we know we have two countries here! But we combine these two on one of our most popular tours, so for us they are one. The vast forests, bogs, fells and fjords make this a magical trip. Owls and woodpeckers feature highly, including magnificent Great Grey Owl and the crow-sized Black Woodpecker. In the Arctic, displaying Ruff, Steller's Eider and Gyr Falcons make a wonderful mix.

Our personal bird highlight: this has got to be Great Grey Owl – it's such an iconic species of these beautiful northern forests.

4. Morocco

Africa meets Europe here and the mix of birds reflects this, with many familiars such as migrating Swallows alongside less familiar Black-crowned Tchagra! The scenery is jaw-dropping from the snow-capped Atlas Mountains to the vast Sahara Desert and the Atlantic coast. In a two-week trip it is possible to enjoy some two hundred species, and delicious food and wonderful hotels add to the enjoyment of this country.

Our personal bird highlight: The brightly coloured, perky Moussier's Redstart, which we saw for the first time high up in the Atlas Mountains.

5. United States of America

A huge country with so many wonderful birds in amazing scenery, it also has very friendly birdwatchers to help you enjoy their bird life. Travelling is easy, with great roads, good food and hotels all adding to the enjoyment. We particularly enjoy birdwatching in Texas in spring, when millions of migrant birds are pouring in from their wintering grounds in South and Central America – high-adrenalin stuff!

Our personal bird highlight: coming face-to-face with Bald Eagle, the iconic national bird, is an experience we won't forget in a hurry.

6. Ecuador

If you want to see a lot of birds, this is the country for you! About the size of Britain and Ireland, it has an incredible 1,600 bird species recorded within its borders. The reason for this is its position on the equator and its amazing diversity of habitats. From the snow-capped peaks of the Andes to the Amazon Rainforest, each life zone has a unique range of species. Ecuador is a modern developed country, making travel easy.

Our personal bird highlight: so many birds to choose from, but having a very close encounter with a Giant Antpitta in the cloud forest puts this bird top of our list.

7. Brazil

A country packed with wonderful birds and home to two of our very favourite places, the Pantanal and the REGUA (Reserva Ecologica de Guapi Assu) World Land Trust reserve. The Pantanal is a vast wetland teeming with birds and wildlife, a paradise on earth. The REGUA reserve is an inspirational conservation project in the Atlantic Rainforest, creating bird- and wildlife-rich habitat from ranch land. Community involvement is key to the success of this project – a model to copy.

Our personal bird highlight: Hyacinth Macaw, a metre-long blue-and-yellow parrot, brought back from the verge of extinction through conservation work in the Pantanal.

8. Namibia

Lying north of South Africa on the Atlantic coast, this huge country is wild and magnificent in equal measure. We love watching birds in big landscapes untouched by man, and Namibia has this combination and more. An amazing variety of habitats, rich in birdlife and complemented by big game, makes for a trip of a lifetime! To watch a huge Martial Eagle in a dead tree with a herd of African elephants drinking from a waterhole below is just heaven for us.

Our personal bird highlight: Ostrich, a bird that's so large we could see it running around on the ground even before our plane had landed in the country!

9. Thailand

Our first visit to this beautiful country was in 2013 and we loved everything about it! The birds were plentiful, tame and stunning, and one of our most-wanted birds was amongst them – the tiny Spoon-billed Sandpiper. This globally endangered species winters here in small numbers and we were thrilled to watch one feeding on a saltpan. Thailand is a picture-postcard country that we will return to for sure.

Our personal bird highlight: seeing Spoon-billed Sandpiper on the saltpans, the bird at the top of our all-time wish list, and one of the most critically endangered species we have been privileged enough to see.

10. Australia

Another vast country, and it has many species of bird found nowhere else on the planet. These endemic species alone make it a must-visit country; add mind-blowing scenery and weird and wonderful animals, and you have a thrilling destination. It was here we broke the world record for the number of species seen in a single year.

Our personal bird highlight: the Bluebonnet Parrot seen on Leeton Golf Course in New South Wales was our species number 3,663 on our Big Year in 2008, thereby breaking the previous world record for species seen within a single calendar year; it's no wonder we have a special love for Australia and its amazing birds.

Bird Observatories in Britain and Ireland

A bird observatory is a centre for studying bird migration and bird populations and, as you can see from the map, they are situated in coastal locations, usually at prime migration points. By carrying out long-term monitoring of bird populations and migration through a daily census, they are able to provide useful information about local birds and migrating species. Many bird observatories trap and ring birds as part of the Bird Observatories Council Scheme coordinated by the BTO, to add to the information we can learn about them. This information is also supplied to national conservation organisations, including Natural England, Natural Resources Wales, Scottish Natural Heritage and their counterparts in Ireland, to help inform their conservation decisions. Some observatories offer accommodation for visiting birdwatchers and volunteers to assist in the ringing, a great chance for a close encounter with exciting birds and a wonderful opportunity to learn more about bird migration first-hand.

Amazing Bird Fact

An Arctic Tern will fly the equivalent of three round trips to the moon over the course of its lifetime.

CHAPTER 6

BIRDS ON THE MOVE

Once you have tasted flight, you will forever walk the earth with your eyes turned skyward, for there you have been, and there you will always long to return.

LEONARDO DA VINCI

Taking Flight

Ask any birdwatcher what particularly attracts them to birds and they'll probably say 'flight'; the fascinating ability to leave the solid ground behind and take to the air; something that we can only achieve with the help of mechanical devices.

The basic mechanics are relatively straightforward. A bird's wing acts as a foil to create lift; air flows over the upper surface of the wing to create an area of pressure lower above than below the wing. Air moves from the area of high to low pressure and that lifts the bird. As the bird flaps its wings, it still creates an area of lower pressure on the upper surface of the wing, again causing lift, but this is now angled forward and it creates thrust to move the bird in that same forward direction.

All flying birds have the same basic structure to their wings to achieve this. The primary feathers are the longest on the outside edge of the wing and provide the forward thrust when flapping; the secondary feathers are shorter and broader and are closely packed to give the bird lift. The tertials are there to cover and protect the primary and secondary flight feathers. The specific quantity and design of feathers and wing shapes are adapted across species to suit how they live, hunt and feed. For example, Hobbies have narrow, pointed wings so they can fly very fast and jink and jive like fighter planes to chase their prey of Swifts and dragonflies. Short, rounded wings allow rapid take-off from a standing start, something that is particularly useful if you are a Pheasant or a Grey Partridge. Broad wings with 'fingers' allow large raptors such as Golden Eagles to soar effortlessly on wind currents. But perhaps the ultimate wing adaptation is that of Hummingbirds, which are able to rotate their wings in a figure-of-eight shape in order to hover on the spot. Sadly, Hummingbirds have never been recorded in the wild in Britain or Ireland; if you think you see one, it will almost certainly be a hummingbird hawkmoth!

Birds with Itchy Feet

In Britain and Ireland, if it's spring, it's the season of northward migration. Many bird species undertake a seasonal journey, often over long distances, between their wintering and breeding grounds. They tend to fly along broad routes known as flyways. Britain and Ireland are situated on a north/south route for many birds that have spent the winter in Africa, and who are moving further north to where they will breed. Some spring migrants may just be passing through Britain on a longer journey to the wide-open spaces of the Arctic tundra, where they will breed and raise their young. But for other species, our woodlands, hills, meadows, uplands, coasts, gardens and eaves are their spring destination. They will find or reclaim a territory, find a new mate or settle down with an existing one, and raise one or more broods of chicks, all in a few short months before making the return journey in the autumn.

Surprising though it may seem, though, we're not just a spring destination. Some bird species move into our country in winter, because our climate here is milder than where they spend the rest of the year. That's all thanks to the North Atlantic Drift, or Gulf Stream, a channel of warmer air and seawater that flows north-east towards our shores, ensuring that even in winter, our temperatures don't usually drop as low as in Continental Europe. So although they may look the same as our residents, some of the birds that visit your birdfeeder in winter, such as Robins or Chaffinches, may in fact be visitors from abroad, escaping the bitter winters of northern Europe. And if you're a Pale-bellied Brent Goose and have spent your summer in Arctic Canada, then winter on the west coast of Britain will feel positively tropical compared to the temperatures you'd experience if you stayed put.

This movement of birds north and south is driven by one thing: food. The drive to find a good food source, either for their chicks in spring or for themselves in winter, is sufficient to make some bird species undertake incredibly long journeys which involve an immense amount of physical effort and huge risks too. Many land-based birds have to fly long distances over open water with nowhere to rest and feed en route, while similarly, water birds flying over land may struggle to find a suitable wet area in which to rest up. Natural predators have long since learned that weary migrants provide a ready source of food. In addition to these natural risks, we humans have added even more threats to the survival of migrant birds by destroying habitat at their stopover points and destinations, adding obstacles such as power lines and wind farms, and by hunting them for sport.

Some humans have been aware of bird migration for centuries, with references to it in Ancient Greek writing and the Bible, although it was misunderstood for a long time (remember Gilbert White dredging those ponds for Swallows submerged in the mud). Even today, with the technological advances available to us, we are still learning a lot more about it.

Nowadays, we understand that birds feel the urge to migrate in response to the change in the length of daylight. The timing of their migration is crucial to coincide with the availability at their destination of suitable food for their chicks. The disruption to the regular seasons that we have experienced lately (known as climate chaos), for example with warm weather arriving either earlier or later than usual and much heavier rainfall totals, can have a disastrous effect on the success of the breeding season. Migrating birds may not be able to advance or delay their arrival to coincide with the time when caterpillars hatch to provide a ready supply of nutritious food for their chicks. As a result, the chicks may not be able to survive any unexpectedly wet or cold weather during their early weeks in the nest.

Studies have shown that birds navigate using the stars and also major landmarks. However, birds do not necessarily take a straight up-and-down route as we humans have assumed. Ringing birds, and attaching tiny satellite trackers to birds, has taught us much more about the routes that birds take and the timings of their migration. For example, thanks to the BTO's project tracking Cuckoos on their migration to and from Africa, we now know that, while some birds do fly in a direct line, others take a much more circuitous route around the Mediterranean. One male was even recorded heading back south in June, much earlier than was originally assumed.

Amazing Bird Fact

There may be a hundred thousand individual Starlings in a massed flock, or 'murmuration'. They make incredible swirling shapes like smoke in the sky as each bird reacts to the seven closest birds around it to change direction or speed.

Amazing Achievements

From a humble human perspective, we find it incredible to contemplate some of the amazing feats of migration that birds are able to achieve on an annual basis. Here are just a few migration heroes:

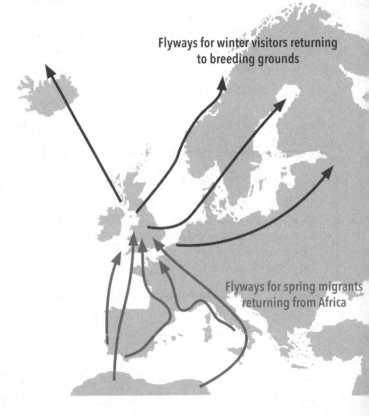

Flyways for winter visitors returning to breeding grounds

Flyways for spring migrants returning from Africa

(Barn) Swallows

Familiar birds which often nest
in close proximity to humans,
such as in the corners of porches and barns, the young will make
their first migration south on their own, completely unaided by their
parents, who will have already left for Africa. The following spring,
these first-summer birds will make their own way back north again
and are likely to return to exactly the same barn or porch where they
were raised.

Arctic Terns

These aerial masters probably see more daylight than any other
creature on earth. Some of these birds spend their winters in the
Antarctic in the southern hemisphere and fly right up north into the
high Arctic to breed. Some do stop
here in Britain to breed, for example
in North Wales, Northumberland
and Scotland.

Bar-tailed Godwits

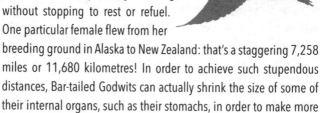

Some of these amazing waders hold the record for the longest migration flight without stopping to rest or refuel. One particular female flew from her breeding ground in Alaska to New Zealand: that's a staggering 7,258 miles or 11,680 kilometres! In order to achieve such stupendous distances, Bar-tailed Godwits can actually shrink the size of some of their internal organs, such as their stomachs, in order to make more space to store fat as fuel for the flight.

Goldcrests

These are our tiniest birds and, as we have seen, they weigh about the same as a 20 pence piece, so it is incredible to think of them migrating. Most of our British Goldcrests don't really migrate, but they do move around to keep ahead of particularly severe winter weather. However, Goldcrests have been recorded as having reached our shores from as far afield as Russia, Finland and Norway.

Vagrants

With all that long-distance travel going on, it is inevitable that some birds will end up in the wrong place by mistake. This may be either because they have been blown off course by strong winds, or because their internal wiring has got a bit mixed up and led them astray. Strong westerly winds may blow American autumn migrants to our shores, particularly on the west coast of Britain and Ireland, while strong easterlies may blow migrants heading for Scandinavia or Siberia onto our eastern coast. Some unlikely arrivals turn up each year and many keen birdwatchers watch the bad-weather forecasts with glee to see what they may bring. Some of our more memorable recent vagrants have included a Common Yellowthroat (a tiny New World warbler) in South Wales; a Sandhill Crane (an American bird that was blown across the Atlantic) in Scotland and then in East Anglia; a White-throated Robin which turned up at Hartlepool instead of Eastern Turkey; a Black Lark which should have been migrating to Kazakhstan but instead found itself on the Isle of Anglesey in north-west Wales; and, perhaps most bizarrely, a Tufted Puffin seen off the Kent coast instead of Canada!

THE CRAZY WORLD OF TWITCHING

When you spend your career in the confines of a gray suit, the Pipits at dawn... are even more wondrous.

MARK OBMASCIK

Twitching is a word often used in connection with birdwatching, but twitching is not watching birds – far from it! Twitching is an extreme type of chasing birds practised by a small minority of birdwatchers. Many birdwatchers would be horrified to be labelled 'twitchers' and, equally, most twitchers would shudder be to called 'birdwatchers'!

So, just what is a twitcher? Simply put, it is someone who travels specifically to see an individual bird, not a species of bird but one particular individual. This will be a bird out of its normal range, a rare bird in the area in which it has been sighted. A twitcher will hear about the sighting and dash to the location to try to see this lost waif and thereby add another bird to their list of birds seen.

Keeping a List

To understand this, we need to understanding 'listing' – keeping a list of birds seen. Listing can take over twitchers' lives; the hard-core twitcher lives for the next addition to his or her list. Twitchers will go to any length to add that new tick to their 'life lists' – a register, if you like, of all the birds they have ever seen. Chartering a plane to reach the Shetland Isles in time to see a Rufous-tailed Robin blown in from Siberia, bunking off work to catch a rarity, abandoning the family mid meal for a rare bird, or, even more extreme, leaving the bride in the lurch – it all goes on in the rather strange world of twitching.

What Makes a Twitcher Tick?

Twitchers come from a surprising variety of backgrounds. You might think this frantic and rather stressful hobby would be the preserve of teenagers, high on energy drinks, but no: middle-aged men, and a very few women, with careers and apparently normal lives undergo a 'Jekyll and Hyde' moment when they hear of a rare bird that would be new to them! They literally twitch with anxiety and stress – will the new bird stay long enough for them to drop everything and dash to the location before it flies away?

The unpredictability of twitching is why people can get hooked on this extreme form of birdwatching. The adrenalin rush on hearing of a rarity, followed by the tension of the race to reach the bird and a second adrenalin rush when the bird (hopefully) comes into view, all combine to make a heady potion! It's a rollercoaster of a ride which can of course end in bitter disappointment if the twitcher arrives too late and the bird has gone. This angst is multiplied if fellow twitchers have arrived earlier and already seen the bird!

With the aid of modern technology, rare-bird news spreads very quickly. Indeed, a rare bird can be seen on a far-flung island and the same day the keenest twitchers will already be arriving by chartered plane. This is not the way that twitching began – far from it! In the early days in the 1960s, things were very different. One of the earliest twitches was to see a Houbara Bustard, an extremely rare bird seen in Suffolk. It was December 1962 when this bird was discovered in a field at Westleton, Suffolk. The few people who saw that rare bird in 1962 heard of its presence by letter! Luckily, the bird liked its Suffolk field and the Royal Mail spread the news in time for the fortunate few.

Why a 'Twitcher'?

It is thought that the label 'twitcher' came about thanks to Howard Medhurst, who was on the first British Rarities Committee in the 1950s. He travelled to see many rarities, riding pillion on his friend Bob Emmett's motorbike. On arriving at their destination, Howard could usually be seen shivering both with cold and the anticipation of an exciting bird to come. This became known as 'going on a twitch' and the label has stuck ever since.

The term 'twitcher' was first used in a publication in 1972 in the British magazine *World of Birds*, and since then 'twitcher' has been mistakenly used to describe anyone with an interest in birds, particularly by the press. For example, on one episode of TV quiz show *Who Wants to Be a Millionaire?* the question, 'Which of these is a popular term for birdwatcher?' was posed. The answers were: A. Twitcher, B. Jerker, C. Blinker, D. Jumper. The contestant was not sure of the answer and opted to ask the audience for help. Incredibly, 93 per cent knew A was the right answer (except, of course, it wasn't!). Twitching had truly come of age!

The Ups and Downs

Twitching can be heartbreaking when things go against the travelling hopeful, as we know from personal experience. Alan once drove through the night from North Wales to Penzance in Cornwall, arriving in time to board the first helicopter over to the Isles of Scilly. Landing safely, he jogged down to the site where the Spotless Starling (the first ever to be recorded in the UK) had been seen. There was the bird sitting on a chimney pot in Hugh Town calling away. It should have been a moment of pure elation after that long and very expensive journey. Sadly, it was quite the opposite. This bird was not a Spotless Starling but an unusually plumaged Common Starling! More than a 1,000-mile (1,600-km) round trip and hundreds of pounds to see a Common Starling – such can be the cruel luck of twitching!

Of course, the good times make the bad times worthwhile. On another occasion in the 1980s, Alan drove to Cornwall to add the Great White Egret to his list, since they were much rarer in those days. Having seen the egret he heard that a Great Spotted Cuckoo had been seen in Kent, so it was back to the car and another overnight drive, this time to Dungeness. At dawn there was no sign of the Cuckoo, but news of a Blue-cheeked Bee-eater in Humberside saw Alan back in the car and driving north! This time the rarity was showing well in the afternoon sunshine, phew! A mad weekend of highs and lows and a lot of driving, over 1,300 miles (2,092 km) clocked up, but two new birds for that all-important life list.

Some of the most famous twitches have involved huge numbers of twitchers descending on an area where a rare bird has been found. A Black Lark on Anglesey in North Wales, all the way from Kazakhstan, was the first to be seen in Britain and Ireland, and it attracted a crowd of over five thousand admirers during its stay! The narrow road to RSPB South Stack, where the bird had taken up temporary residence, was jammed solid with abandoned cars as twitchers sprinted to the heath where the lark was feeding. It was chaos.

One of the earliest mass twitches was to see a Golden-winged Warbler in Kent. It was another first record for Britain and Ireland, so every twitcher needed to see this bird which had taken up residence in a housing estate with lots of cul-de-sacs. The American warbler happily flitted from garden to garden over the fences, but of course the twitchers reaching the end of a cul-de-sac had to turn around and make their way round to the next road. With such a mobile bird and hundreds of people following it around the estate, you can perhaps imagine the chaos. Those people at the front of the queue who had already seen the bird were heading away, while those just arriving at the end of the queue were still surging forward: cue for a head-on collision on foot! Madness! A garden wall was demolished, roads were blocked, residents were unable to get to or from their houses, and the school bus needed a police escort to get through the crowd! This is twitching.

Amazing Bird Fact

Hummingbirds beat their wings up to 50 times per second, the fastest of any bird on the planet.

COMPETITIVE BIRDWATCHING

Veni, vidi, vici. (I came, I saw, I conquered.)
Julius Caesar

Can birdwatching really be competitive? Oh, yes! We have mentioned twitchers and those all-important life lists that they keep, a personal list of all the birds they have seen. Well, once you have a list of your own, it is only natural to compare your list with other twitchers' lists. As soon as you do that, you have a competition. Everyone who is inclined to keep a list of something wants it to be a long list; this is natural behaviour, so all twitchers want to have a long list of birds they have ticked off. They also want their list to be longer than other twitchers' lists to give them status amongst their peers. So, the race is on: who can see the most birds and have the biggest list?

What's in a List?

But it does not stop there, oh no! There are lots of lists that twitchers and also some birdwatchers keep. Perhaps the most important to many people is their 'life list'. As the name suggests, this is a catalogue of all the species of bird seen during that person's life. However, for many British and Irish twitchers, their 'country life list' is the most important, in other words, all the birds seen in the UK or Ireland during that person's life. Websites such as www.Surfbirds.com publish the details of these life lists so that competitive birdwatchers can see how they personally are doing compared to others. This drives the competition, of course. If a twitcher sees that he or she is just behind another, they may be spurred on to try harder, spend more money, travel more miles, or bunk off work more often in order to move up the league table. Of course, with the life-list challenge, age is an advantage. If you started your twitching 'career' back as a young teenage birder, for example back in the 1960s or 1970s, and have maintained the enthusiasm continuously since then, your life list will be very long indeed. A newcomer to the twitching scene can only dream of all those amazing birds under the older twitchers' belt. So the life-list game is a long one, and of course, the longer it goes on, the harder it becomes (and the further you may have to travel, incurring more expense) to find a new bird to add to your list.

That brings us nicely to the 'year list'. As you may have guessed, this is a list of birds seen in one calendar year. Now this is a much more competitive race as the slate is wiped clean at midnight on 31 December each year and a new race begins on New Year's Day. This unofficial race is open to all mad enough to enter. Many people set a geographical limit to the challenge, usually the country they live in, so they will have a UK or Irish year list. It is great fun, if you don't take it too seriously, and a nice way to compare one year's birdwatching with another. However: beware! These things can get out of hand. One UK twitcher called Lee Evans has undertaken a fanatical UK year list every single year since the 1970s! Lee has certainly travelled more miles and seen more individual rare birds in the UK than anyone else, and has often won the Year List race. Anyone taking on Lee for this particular race needs to have planned well, banked a lot of money

and have tremendous energy to keep going flat out all year to see around 380 species. To put that number in context, an average keen birdwatcher might see around 225 species in a year, so you can see that a huge amount of twitching is needed to win.

Year lists and life lists are taken largely on trust. It is down to the individual to decide for themselves whether he or she has in fact seen a particular bird and whether they feel comfortable enough with the sighting to put it on their own list. Ask most birdwatchers and they'll tell you that it just doesn't feel right to put any 'dodgy' or dubious sightings on their list!

Across the Pond

If you think doing a year list in the UK or Ireland is mad, then how about doing one in North America? The distances are huge and the amount of travel required is mind-blowing! Back in the 1970s, a very young American birdwatcher, Kenn Kaufman, dropped out of school with one aim: to set a new North American year-list record. No one would have given him a chance, but Kenn was made of tough stuff and he set off hitchhiking, criss-crossing his way across the USA, often sleeping rough, living on an average of just one dollar a day. His grit and single-minded determination carried him through 365 days of non-stop chasing and he set a new record with 671 species. This record has since been beaten many times, but never by anyone with a budget as restricted as Kenn's.

> *We were out to seek, to discover, to chase, to learn, to find as many different kinds of birds as possible – and, in friendly competition, to try to find more of them than the next birder.*
> KENN KAUFMAN

Amazing Bird Fact

Members of the crow family have the largest brains relative to their body size of all bird species.

If the United States isn't big enough to give you enough birds, how about setting a world birdwatching record. As you can imagine, this requires a huge amount of planning, travel, dedication and a healthy dose of good luck too. As a result, there have not been many contenders.

World record holder	Year	Total
James Clements, USA	1989	3,662 species
Alan Davies and Ruth Miller	2008	4,341 species

The Biggest Twitch

Back in 1989 an American birdwatcher named Jim Clements set a new world record by logging an amazing 3,662 species of bird in a year of crazy travel. This record stood for a long time, which is hardly surprising given the complexities and sheer madness of taking on a global world record. Then, in late 2007, two British birdwatchers announced that they were going to try to beat this world record of 3,662 species. Those two birdwatchers were Alan Davies and Ruth Miller, your authors.

In 2008, we set off on a non-stop year of birds and travel in pursuit of that world record, visiting 27 countries over the course of 12 months! It was fast, mad, exhausting and the best thing we have ever done. Our year began on New Year's Day in Arizona at a sewage works, where bird number one was the Cactus Wren, the state bird of Arizona. From the US we headed south to Mexico and then into Ecuador, where we dashed to over a thousand species by the end of January!

By October we were in Australia and closing in fast on that world record. With the help of our great friend Iain Campbell, we set a crazy pace Down Under and on the last day of the month we went to a golf course. Iain knew golfers like their greens to be green, so they water them well; in the dry outback water pulls in birds, so perhaps we'd be lucky here. We lifted our binoculars and focused on a gang of parrots that had flown in to take advantage of all that greenness. They were Bluebonnet Parrots, a new species for our year list, but not just any old bird; they were bird number 3,663! That new world record for the number of bird species recorded in a single calendar year was ours, and we were elated.

With two months of the year still to go we did not stop, however. We briefly paused to open a bottle of champagne, before heading out to look for more birds, and by 31 December we had amassed a total of 4,341 species – that is nearly half the world's birds in just one year. It was a year of amazing birds in wonderful places with plenty of adventures which included being held up at gunpoint, nearly drowning in a sinking boat, being robbed, and meeting so many great people! Our book, *The Biggest Twitch*, documents the warts-and-all story of our Big Year.

If devoting a whole year to competitive birdwatching sounds a little daunting, then you could always opt for one day. A bird race is a popular thing that many birdwatchers enjoy and, as the name suggests, it is a race to see as many birds as possible but with a strict time limit, usually 24 hours. Again, geographical limits can be set, so a county or region can be the field of play. In the USA they even hold the World Series of Birding! It is open to all nationalities, though the teams are usually mainly US-based. It takes place in the state of New Jersey each May and teams have 24 hours to record as many species in the state as possible. This is taken very seriously, with teams planning their routes weeks in advance with split-second timing to maximise bird numbers and minimise time spent between new species. To win, you need determination and the ability to stay awake and enthusiastic for the whole 24 hours!

But a new day-list gauntlet has been thrown down. A team from Louisiana State University set a new Big Birding Day record on 14 October 2014, seeing an incredible 354 bird species in just one day in Peru. That's more birds than most birdwatchers can see in a lifetime in Britain and Ireland!

Dr Mark Avery was born in 1958 and has been passionate about birds and wildlife since his early years as a member of his school's Field Club. After studying science at university, Mark joined the RSPB staff in the mid 1980s, initially on a two-year contract. Over a 25-year career Mark rose to become Director of Conservation, a role he held for nearly 13 years. During his time with the RSPB, he tackled the thorny issues that the society faced, ranging from direct conservation activities to political lobbying. He left the RSPB to become a freelance writer, broadcaster, lobbyist and conservation consultant, and he has been particularly active in taking a stand against the illegal killing of Hen Harriers on managed grouse moors.

CHAPTER 9

BIRDS AND MAN

*Everyone likes birds. What wild creature is
more accessible to our eyes and ears?*

DAVID ATTENBOROUGH

Part 1: Birds in Ancient Culture and Folklore

Birds have been a feature of human culture for at least as long as
man has been able to draw or write. From the time of the earliest
cave paintings onwards, we have been fascinated by birds and
their ability to fly. In Northern Australia, palaeontologists found a
rock painting which depicted a bird that became extinct over forty
thousand years ago.

Amazing Bird Fact

*The chicken is the closest living relative to
Tyrannosaurus rex.*

Birds in Mythology

Some of the most important gods of Ancient Egyptian mythology bear bird features. Horus, a man with a falcon's head, is one of the oldest and most important, being the god of the sun, the moon, war and protection. The heron was another deity, known as Bennu, the god of the sun, creation and rebirth, and may have been the forerunner of the mythical phoenix, a bird which burned to death after living for 500 years, only to be reborn as a new bird from its own ashes. Thoth was the god of wisdom, learning and writing, and is depicted with the head of an ibis, while Nekhbet was seen as the white-vulture goddess and protector of Upper Egypt.

According to Greek mythology, harpies were female monsters with the winged body of a bird and the face of a woman. Considered evil characters, they were wind spirits, notorious for stealing food from their victims just as they were about to eat. The Greeks also believed that the virgin goddess of wisdom, Athena, was always accompanied by a Little Owl. This partnership also continued into Roman mythology, where the Little Owl played sidekick to Minerva. Since then, owls have always been associated with knowledge and wisdom in the western world.

Birds in the Good Book

Birds appeared in the Bible too. According to the Book of Genesis, after joining Noah in his ark with all the other animals being loaded in two by two, birds played a leading role when the floodwaters started to recede. Although the raven flew off without returning, the dove returned bearing an olive leaf, proof that the flood levels had dropped and that there was land and safety nearby. And in the New Testament, Jesus told his disciple Peter that he would disown him three times before the cock crowed. Birds had an important role to play in Roman religion too, as priests would advise on the course of action to be taken by interpreting the words of birds or watching their activities to predict future events.

Birds in Folklore

How we perceive birds and what we associate with them vary across different countries and cultures. Owls are still associated with learning and wisdom in Europe. As the verse goes:

> *A wise old Owl sat on an oak,*
> *The more he saw the less he spoke,*
> *The less he spoke the more he heard,*
> *Why aren't we like that wise old bird?*

But in parts of Africa owls have long been associated with bad luck, witchcraft and death. For example, in East Africa, the Kikuyu tribespeople in Kenya are said to believe that owls will bring illness and death to children, while the Zulus of South Africa believe that they act as messengers for wizards and witches. Even today, owls, particularly Barn Owls, are often killed in West African countries as they are thought to bring bad luck. Conversely, in Ancient Egypt and Persia, Hoopoes were considered sacred and virtuous, while in Europe they were associated with thieving and even with being harbingers of war in Scandinavia. Ravens have had a pretty bad press everywhere, being associated with ill omen, death and destruction, perhaps because they would often be found taking advantage of the carnage of a bloody battlefield. Flying in the face of this, however, the Raven is the official bird of the Yukon Territory and the city of Yellowknife in Canada. Storks are universally positive characters across cultures and religions, being seen as the very model of parental devotion and famous, of course, for delivering babies.

Amazing Bird Fact

Ravens can make up to 30 different kinds of sound or vocalisation.

Folk Names of British Birds

Before birdwatching became formalised with Latin names for every species, everyday people had their own idiosyncratic ways of describing what they saw. This list, taken from a book published in 1913, compiled by H. Kirke Swann, gives a few of the more interesting ones:

Ailsa Parrot: local Scots name for the Puffin.

Apple-shealer or Apple-sheiler: Northumbrian names for the Chaffinch, from the bird's habits among the buds of fruit trees.

Greedy Glead: the Kite.

Green Grosbeak: the Greenfinch.

Isle of Wight Parson: the Cormorant (Hampshire).

Jetcock: the Jack Snipe.

Leg Bird: a provincial name for the Sedge Warbler.

Links Goose: the common shelduck, for its habit of frequenting the 'links' (sandy plains) near the sea in Scotland.

Mavis: the Song Thrush. Possibly from the French *mauvis* (redwing).

Merwys: poetical Welsh name for the Blackbird.

Our Lady's Hen: a Scots name for the Wren.

Peep o' Day: the Little Grebe.

Row-dow or Roo-doo: the House Sparrow.

Sallypecker: the Willow Warbler (Ireland). 'Sally' signifies 'sallow' (willow).

Scissors-grinder: the Nightjar, from its jarring note.

Part 2: Birds in Modern Culture

As well as being integral to ancient culture, birds have featured in more modern culture too, having been an inspiration in poetry, literature, art and film.

Birds in Poetry

A poem written in Persia in 1177 called 'The Conference of the Birds' tells the tale of how the birds of the world gathered together to decide who should be their king and help them find enlightenment:

Once on a time from all the Circles seven
Between the steadfast Earth and rolling Heaven
THE BIRDS, of all Note, Plumage, and Degree,
That float in Air, and roost upon the Tree;
And they that from the Waters snatch their Meat,
And they that scour the Desert with long Feet;
Birds of all Natures, known or not to Man,
Flock'd from all Quarters into full Divan,
On no less solemn business than to find
Or choose, a Sultan Khalif of their kind,
For whom, if never theirs, or lost, they pined.

The Hoopoe, considered the wisest amongst them, suggests they visit the Simorgh, a phoenix-like mythical bird, who will be able to advise them. When they get there, they find only a calm lake which reflects their own images and imperfections, suggesting that the path to enlightenment is internal, not external.

Geoffrey Chaucer wrote his poem 'The Parlement of Foules' (The Parliament of Fowls) in the fifteenth century, telling the tale of a dream about Nature convening a meeting where all birds were to choose their mates. While the small birds mate, the female falcon is allowed to delay her decision for a year, perhaps an allegory for freedom of choice!

> A gardyn saw I, ful of blosmy bowes
> Upon a ryver, in a grene mede,
> There as swetnesse everemore inow is,
> With floures whyte, blewe, yelwe, and rede,
> And colde welle-stremes, nothyng dede,
> That swymmen ful of smale fishes lighte,
> With fynnes rede and skales sylver bryghte.
>
> On every bow the bryddes herde I synge,
> With voys of aungel in here armonye;
> Some busied hem hir bryddes forth to brynge;
> The litel conyes to here pley gonne hye.
> And ferther al aboute I gan espye
> The dredful ro, the buk, the hert and hynde,
> Squyrels, and bestes smale of gentil kynde.

Shakespeare makes more reference to birds in his works than any other writer of Western literature. He frequently refers to trapped or snared birds as representing the worst example of terror that a creature could feel. In contrast, he makes many references to swans in a positive context, for example in *Henry VI Part I* he writes:

An earl I am, and Suffolk am I call'd.
Be not offended, nature's miracle,
Thou art allotted to be ta'en by me:
So doth the swan her downy cygnets save,
Keeping them prisoner underneath her wings.
Yet, if this servile usage once offend.
Go, and be free again, as Suffolk's friend.

Samuel Taylor Coleridge refers to an albatross in his poem 'The Rime of the Ancient Mariner', published in 1798. His eponymous mariner shoots an albatross as it follows his ship; usually a sign of good fortune. After the killing his ship suffers terrible misfortunes. The mariner feels that this is all his fault and that the albatross is a metaphor for the curse or burden that has been put upon him:

And I had done a hellish thing,
And it would work 'em woe:
For all averred, I had killed the bird
That made the breeze to blow.
Ah wretch! said they, the bird to slay,
That made the breeze to blow!

The Victorian novelist and poet George Meredith wrote about the Skylark in a poem that would go on to be the inspiration for Vaughan Williams' musical masterpiece (see below), perhaps himself inspired by hearing it sing in flight over Boxhill in Surrey, where he lived:

He rises and begins to round,
He drops the silver chain of sound
Of many links without a break,
In chirrup, whistle, slur and shake,
All intervolv'd and spreading wide,
Like water-dimples down a tide
Where ripple ripple overcurls
And eddy into eddy whirls.

More recently, in 2008, poet Paul Farley wrote 'For the House Sparrow, in Decline'. In it he describes the role that the common (though, sadly, increasingly less so) House Sparrow plays in our modern lives.

Perhaps more light-hearted is Edward Lear's famous children's poem 'The Owl and the Pussycat', about an unlikely pair of lovers:

The Owl and the Pussy-cat went to sea
In a beautiful pea-green boat,
They took some honey, and plenty of money,
Wrapped up in a five-pound note.
The Owl looked up to the stars above,
And sang to a small guitar,
'O lovely Pussy! O Pussy my love,
What a beautiful Pussy you are,
You are,
You are!
What a beautiful Pussy you are!'

Birds in Literature

A quick look at our own bookcases shows that there are plenty of factual books about birds – field guides to the birds in a specific country, a focus on the birds found in particular counties, monographs of individual bird species, summaries of rare birds to visit Britain and Ireland… the list is endless. There are countless books about birders too. Many of these have a common theme: what birds can be enjoyed given specific time or geographic restrictions. *A Twitcher's Diary* (once blacklisted by the RSPB) shows Richard Millington's bird notes and sketches, particularly of rarities, of a year of birding in the UK in 1980. *Kingbird Highway* tells the tale of Kenn Kaufman, the young birder who criss-crossed the United States in 1973 on a shoestring (including surviving on tins of pet food!) to see as many American birds as possible in a year. *The Profit of Birding* by British birding legend Bryan Bland is the tale of his exploits over a lifetime of birdwatching tours around the world. *The Urban Birder* is the story of David Lindo and where his passion for birding took him – from urban east London to Hollywood and beyond. And maybe our own book *The Biggest Twitch* deserves a place on your bookshelf, telling the tale of how we gave up our jobs and sold our home to travel around the world for a year to set a new birdwatching world record.

But do birds also feature in modern fiction? Rabbits have their *Watership Down* and otters have *Tarka*, but what about birds? They do play significant parts in some children's literature, particularly owls. Owl is a key character in A. A. Milne's books about Winnie the Pooh, and they also feature regularly in the Harry Potter novels. The book called *A Kestrel for a Knave* by Barry Hines is perhaps better known as the film entitled *Kes* about the relationship between a young boy and the Kestrel that he trains. But there aren't that many fictional bird stars, so clearly there is an opening here for budding authors to redress the wildlife balance.

Iolo Williams was born in Wales in 1962 and is a passionate naturalist, conservationist and birder. He worked for 14 years for the RSPB in the field, monitoring rare birds like the Red Kite, Chough and Black Grouse, and as a regional coordinator, but he is perhaps best known for his wildlife programmes on television. He has featured in Welsh-language programmes for S4C, such as *Crwydro* (Roaming), and in English-language programmes for the BBC, including *Wild Wales* and *Rugged Wales*, and has done much to raise the awareness of the beautiful Welsh countryside amongst a wider audience. Iolo has also made increasingly regular appearances in the BBC *Springwatch* programmes.

Birds in Art

John James Audubon, Roger Tory Peterson and Sir Peter Scott have played a significant role in developing bird art, as we have already seen. Other, more modern, notable bird artists include:

- **Charles Frederick Tunnicliffe** (1901–1979) is known for his realistic wildlife paintings, mainly in watercolour and oil. A prolific artist, he illustrated the cover of Henry Williamson's book *Tarka the Otter*, as well as many of the Ladybird bird and wildlife books that some wildlife enthusiasts, including the two of us, were brought up on. He also illustrated Brooke Bond tea cards and covers for the RSPB members' magazines called *Bird Notes* and *Birds*. Many of his paintings are of the waders and wildfowl he could see from his home overlooking the Malltraeth marsh in North Wales, and some of his works are on exhibition at the Oriel Ynys Môn Anglesey Heritage Gallery at nearby Llangefni. www.thecharlestunnicliffesociety.co.uk

- **Robert Gillmor** (1936–present) is a founder member of the Society of Wildlife Artists and is particularly known for his brilliant and dynamic lino-cut illustrations of birds. He has been illustrating the *New Naturalist* book jackets since 1985. www. swla.co.uk

- **Killian Mullarney** (1958–present) is an Irish ornithologist and bird artist who creates exquisitely detailed and accurate illustrations of birds. He is perhaps most famous for illustrating, together with Swedish illustrator Dan Zetterström, the *Collins Bird Guide* by Lars Svensson and Peter J. Grant, considered to be perhaps the definitive guide to the birds of Britain and Europe.

To get a better idea of the wealth of bird artists that there are in Britain, why not visit an exhibition by the Society of Wildlife Artists, or drop into the Art Marquee at the British Birdwatching Fair at Rutland Water in August each year, where artists display and sell their bird art.

Birds in Classical Music

It seems natural that melodic birdsong should also find its way into classical music. George Meredith's poem about the Skylark (see above) inspired Ralph Vaughan Williams to compose his orchestral masterpiece, *The Lark Ascending*, where the soaring notes from the violin evoke the Skylark's song from high in the sky on a warm summer's afternoon.

And in a brave experiment, when the BBC broadcast its first live outdoor performance in 1923, it was a concert featuring an unlikely duet between a cellist and live Nightingales. This incredible broadcasting achievement was recreated in 2013 at RSPB Northward Hill in Kent, featuring cellist Clare Deniz.

Here are some other examples of classical composers including birds in their work:

Beethoven *Symphony 6 Pastoral*, Second Movement (includes Cuckoo, Nightingale, Quail)

Delius *On Hearing the First Cuckoo in Spring*

Handel *Concerto for Organ in F Major, The Cuckoo and the Nightingale*

Haydn *Quartet for Strings Op. 64 No 5, The Lark*

Mozart, Leopold *Toy Symphony* (Cuckoo)

Respighi *Gli uccelli (The Birds)* (doves, hens, Nightingale, Cuckoo)

Rimsky-Korsakov *Invisible City of Kitezh* (Cuckoo)

Rimsky-Korsakov *The Snow Maiden* (Cuckoo)

Saint-Saëns *Carnival of the Animals* (Cuckoo, hens, roosters)

Strauss, Johann Jr *Nightingale Polka Op. 222*

Strauss, Johann Jr. *Im Krapfenwald'l Polka Op. 336* (Cuckoo and other bird sounds)

Strauss, Josef *Village Swallows Waltz Op. 164*

Vivaldi *Violin Concerto in A, The Cuckoo*

Birds in Pop Music

Birds have also worked their way into our psyche through pop songs. From Vera Lynn's wartime rendition of 'There'll be Bluebirds Over the White Cliffs of Dover' to 'Birds of the High Arctic' by David Gray, birds have inspired singers and songwriters over the years. Here are a few more examples:

- 'When the Red, Red, Robin Comes Bob, Bob, Bobbin' along' – Al Jolson, 1926
- 'The Ugly Duckling' – Danny Kaye, 1952
- 'Blackbird' – The Beatles, 1968
- 'Albatross' – Fleetwood Mac, 1969
- 'Ride a White Swan' – T. Rex, 1970
- 'Freebird' – Lynyrd Skynyrd, 1974
- 'Fly Like an Eagle' – Steve Miller Band, 1976
- 'Bluebird' – Electric Light Orchestra, 1983
- 'Dodo' – David Bowie, 1990
- 'Mocking Bird' – Barclay James Harvest, 1997
- 'Starlings' – Elbow, 2008
- 'Morning Mr Magpie' – Radiohead, 2011

Perhaps the ultimate 'birdy' album is Avocet *released by the Scottish folk musician Bert Jansch in 1979. All the tracks are instrumental and named after birds:*

Avocet
Lapwing
Bittern
Kingfisher
Osprey
Kittiwake

Pop Group Names with Bird Connections

Some birds also lend themselves to the names of pop groups. Here are some fine examples:

Guillemots: English four-piece indie band hailing from Birmingham.

Eagles: the American rock band best remembered for 'Hotel California'.

A Flock of Seagulls: English new-wave band from Liverpool.

Sheryl Crow: eclectic American singer/songwriter and guitarist.

The Black Crowes: American rock band formed in Georgia.

Counting Crows: American rock band from California.

The Housemartins: English alternative rock band.

The Yardbirds: English rock band that originated in the sixties.

Birds on the Silver Screen

Birds do appear on the big screen, though sometimes this has more to do with the name than the content of the film:

The Maltese Falcon (1941): Sam Spade (Humphrey Bogart), a private detective from San Francisco, foils attempts to steal the bejewelled statue of a falcon.

To Kill a Mockingbird (1962): Based on Harper Lee's award-winning book about how lawyer Atticus Finch (played by the appropriately named Gregory Peck) deals with crime and racial inequality in the American Deep South.

Where Eagles Dare (1968): A World War Two movie set in Bavaria as the Allies try to rescue an American Brigadier General captured by the Germans.

A Nightingale Sang in Berkeley Square (1979): Convicted criminal Pinky (Richard Jordan) takes a job at a large bank, putting temptation in the way of himself and local hoodlum 'Ivan the Terrible'.

The Pelican Brief (1993): Law student Darby Shaw (Julia Roberts) uncovers a lethal plot to develop protected land and manipulate the judiciary in favour of the developer.

Birds do sometimes star in the movies themselves, as in Alfred Hitchcock's 1963 suspense film *The Birds*. Set in Bodega Bay in California, it tells of citizens who were subjected to horrifying attacks by birds, which didn't exactly give birds and the enjoyable pastime of birdwatching a good press.

However, 2011 saw the release of a film called *The Big Year*, which was actually about birdwatching and went a long way to redress the balance about the nature of birds. It was (loosely) based on the non-fiction book of the same name by Mark Obmascik about three Americans competing to see the highest number of bird species in North America in a single year. The film starred Steve Martin, Jack Black and Owen Wilson and took a comedy approach to their Big Year, although the actual event was taken very seriously indeed by the original participants.

Amazing Bird Fact

The Ostrich lays the largest egg in the world. It is the size of a cantaloupe melon – ouch!

Birds on the Beeb

Britain has a wonderful tradition of broadcasting wildlife programmes on national television. From the time *Zoo Quest* (the first wildlife programmes to feature Sir David Attenborough) started in the 1950s, followed by Johnny Morris's children's programme *Animal Magic* from the 1960s onwards, we have become accustomed to having amazing footage of wildlife being brought into our living rooms. As the years have passed, so our knowledge of birds and wildlife worldwide has increased (along with a growing degree of sophistication in the photography of them). Birds were given their own complete BBC series in *The Life of Birds* in 1998, which covered every aspect of birds' lives. *Birds Britannia in 2010* took us through the relationship between birds and people in Britain, and in 2011, *Earthflight* allowed us to take virtual flight ourselves and experience what it is like to see the world on the wing. But don't be fooled by one title: the long-running *Birds of a Feather* (1989–2014) has a lot more to do with prison than flight!

Tweet of the Day has recently graced our radios too, via Radio 4 (2014); nothing to do with birds on social media but a daily short broadcast of bird songs, first within Britain and Ireland and then moving on to more exotic species overseas. What a great way to start your day if you can't get out to hear the real thing.

CHAPTER 10

EXTINCTIONS AND INTRODUCTIONS

*In pushing other species to extinction, humanity is
busy sawing off the limb on which it is perched.*

Paul Ehrlich

Populations of bird species around the world are not static, and as
we continue to study them, so we are learning more about them.
Thanks to superior scientific knowledge and methods of testing, such
as studying the DNA of birds, we now know that some that were once
considered separate species are now in fact the same species. This is
called 'lumping' in taxonomic terms. Equally, there are other species
that used to be thought all one and the same, but which have now
been found to be separate species. This is called 'splitting'. In addition
to changes in how we classify birds, we know that some species have
become totally extinct anywhere in the world, while others have
found their way, usually with the aid of man, to completely new parts
of the world.

Extinctions

Since bird species were first scientifically named and noted in the 1500s, over 190 known separate species around the world have become extinct. Sadly, this rate of extinction seems to be increasing. There are around ten thousand species of birds worldwide, and approximately 1,200 of these are under threat of extinction. In the UK, we have a red/amber/green system for categorising birds by conservation importance; red is the highest risk status, meaning the need for urgent conservation action. Amber is the next critical level, while species with a green status are deemed not to be at risk. Birds in Britain are monitored by a collaboration of scientific and conservation organisations including the British Trust for Ornithology and the RSPB, so that we remain aware of any declines in species populations and can take action to reduce the threats. A detailed report on the situation is provided by the RSPB on behalf of a number of NGOs involved in the study. The latest edition is entitled 'The State of the UK's birds 2014', the 15th edition of this annual report, and it is available as a PDF download from the RSPB website on www.rspb. org.uk/forprofessionals/science/sotukb.

However, in the case of some species, this is all too late and they have vanished from our planet forever. Here are just a few examples.

The Dodo

The Dodo is perhaps the best-known extinct species and, as such, is often used as a symbol of extinction or obsolescence. It had the misfortune to be both discovered and wiped off the face of the earth within the space of a hundred years, and this drew attention to the fact that humans could be the cause of other species becoming extinct.

The Dodo was found only on the island of Mauritius, where, with wings too small for flight and laying a clutch of just a single egg, it survived because of an abundance of food and a lack of predators. Unfortunately for the Dodo, visiting sailors in the sixteenth century quickly realised its potential as an easily caught nutritious meal and the poor bird was unable to withstand the combined onslaught of hunting by humans and introduced predators. The last live Dodo was recorded in 1662. Fossil remains suggest it was 3.3 feet (1 m) tall and weighed 22–40 pounds (10–18 kg), but with only a few drawings made during its lifetime, we're not entirely sure exactly what it looked

like. However, we all have a pretty good idea of how it might have looked, thanks to Lewis Carroll and the Dodo's appearance in his 1865 book *Alice in Wonderland*.

The Elephant Bird

Elephant Birds were giant flightless birds that used to exist on Madagascar. Although they looked like Ostriches, they were in fact more closely related to Kiwis. They became extinct in the seventeenth century, probably thanks to human intervention. Their vital statistics are staggering: they were believed to have been the world's largest bird at the time, standing over 10 feet (3 m) tall and weighing nearly 880 pounds (400 kg), and laying an egg of over 3 feet 3 inches (1 m) in circumference and 13 inches (34 cm) long – ouch again!

The Passenger Pigeon

In 2014 the Passenger Pigeon celebrated, if that is the appropriate word, the centenary of the death of the last individual of the species. She was known as Martha and she lived and died in captivity in Cincinnati Zoo, Ohio. This species, perhaps more than any other, teaches us a hard lesson in wildlife conservation. The pigeon gets its name from the French word *passager*, which means 'to pass by'. This reflects its one-time status as the most abundant bird in America, when it represented nearly one in every four birds in the entire country. It is said that the skies went dark as huge flocks of Passenger Pigeons passed overhead; one flock was reported to take 14 hours to fly by as over 3.5 billion birds flew in a flock that was one mile (1.6 kilometres) wide and 300 miles (483 kilometres) long, a sight that is impossible to imagine now. Thanks to humans, the Passenger Pigeon went from being plentiful in the 1800s to being extinct by 1914.

First, the early European settlers deforested large areas of land, which destroyed much of the key habitat for the pigeon. Then its meat became popular as a cheap food source for slaves and poor freemen, leading to hunting on an industrial scale. With flocks containing so many individual birds, it must have seemed that there was an endless supply of Passenger Pigeons to use for human requirements. But with slaughter on such an enormous scale, the bird quickly went into decline. Attempts to keep the species alive in captivity failed and the Passenger Pigeon became extinct.

The Great Auk

The Great Auk was part of the alcid family that includes Razorbills and Guillemots, which you can see breeding at coastal sites in Britain and Ireland. However, unlike both of these birds, the Great Auk could not fly, although, like penguins, it could swim well underwater. It was a big bird at around 33 inches (84 cm) tall and it weighed 11 pounds (5 kg). It bred on isolated islands, but it hunted for fish across a huge ocean range from the coast of Canada to Greenland, Great Britain and as far south as northern Spain.

The bird was hunted both as a source of food and as a totemic symbol in North America, and its feathers were sought-after as a supply of warm down in Europe. Scientists realised that its numbers were falling dramatically and attempts were made to halt its decline, though its rarity factor only added to the appeal of the skins and eggs of the Great Auk in the eyes of European collectors. As a result, the last two known birds were killed in 1844 off the coast of Iceland.

The Ivory-billed Woodpecker

The Ivory-billed Woodpecker was one of the largest woodpeckers in the world and it lived in the swamps and pine forests of south-east America, where its distinctive 'tock-tock' drum would resound amongst the trees. These large birds, measuring around 20 inches (51 cm) long with a wingspan of 30 inches (76 cm), needed large tracts of wet forest in which to live, but logging on an industrial scale destroyed their habitat to such an extent that it is unlikely that there are any forests left that are large enough to support the Ivory-bill.

Its exact status is unconfirmed and it is classified as 'definitely or probably extinct'. However, there is still some uncertainty and there were reports of a male bird being seen in Arkansas in 2004. The bird has never been re-located despite intense searching, and even the offer of a $50,000 reward for anyone who can lead the research team to a living example has failed to produce an Ivory-billed Woodpecker. Conservation efforts have been made to protect suitable habitat in areas where individuals may still be hanging on, so what are you waiting for? That reward is still up for grabs!

Amazing Bird Fact

A record-breaking Puffin has been logged carrying 61 sand eels in its gaudy bill at a single go. That's one lucky puffling in the nest who didn't go hungry!

Introduced Species

Some of the birds you may see living in the British countryside are not, in fact, native. They may have been specifically introduced for one reason or another, perhaps for food, or they may have escaped from private collections and continued to live successfully in the wild. In some cases, they look so much at home it's as if they've always been there, while other species definitely add a touch of the exotic to our landscapes.

Common Pheasant

The Common or Ring-necked Pheasant is native to Asia but was first introduced to Britain in the tenth century as a source of food. It's been here for so long that you would be forgiven for thinking it is truly native and, introduced or not, it is undeniably a handsome bird. If you look closely you will see quite a range of colour schemes. Some birds are the more 'traditional' russet colour, some are almost white, while others are a smart bottle-green colour, as a result of captive breeding and hybridisation over the centuries.

Pheasants slipped under the radar for a long time but they were rediscovered as a game bird in the 1800s. Since then they have been reared in huge numbers by gamekeepers with up to 30 million birds being released on shooting estates each year. Most released birds are killed within the first year either through shooting or their own lack of road sense, but some birds do survive in the wild and they have a widespread distribution across the country.

Canada Goose

This species looks so much at home here that again you might think it is native to Britain but, as its name suggests, it comes originally from Canada and the northern United States. On its home turf, the Canada Goose has suffered quite a significant decline thanks to overhunting and habitat destruction, but here in Britain the living is easy and the bird is thriving.

Although some birds have found their own way to Europe through natural range expansion, Canada Geese were first introduced into Britain in the seventeenth century as part of King James II's private wildfowl collection in St James's Park in London. They were also introduced across Britain by hunters, but for a long time their numbers never really took off. However, if we fast-forward to the twentieth century, Canada Geese have recently undergone a rapid multiplication and the BTO estimates that we currently have an overwintering population of approximately 190,000 birds. The adult birds have no natural predators, though a hungry fox would happily take a chick given the chance. We have plenty of suitable habitat for them, with our parks, golf courses, lakes and nature reserves providing safe water near a suitable food supply. Canada Geese have taken full advantage of this natural abundance and some people may consider them a pest as their droppings contaminate water and land, while their loud noise and aggressive behaviour may make them unpopular.

Little Owl

A bird that's much harder to see here than the ubiquitous Canada Goose is the Little Owl. This small but feisty bird is native to much of southern Europe, North Africa and parts of eastern Asia. It was first introduced into Britain by Thomas Powys, the Fourth Earl of Lilford, who was a keen ornithologist and bird collector with a private collection that included two free-flying Lammergeiers (Bearded Vultures)! The Little Owl has naturalised across Britain, where it lives in open farmland and parkland. As it is partly diurnal, you may see it perching on a regular spot on a rock or tree branch. It feeds primarily on insects and earthworms. However, although it is only 8 inches (20 cm) tall, it will also eat small mammals and birds and quite happily tackle prey much larger than itself, such as grouse. While non-native introductions are not necessarily approved of today, the Little Owl has not had a negative impact on the British countryside and, like most owls, has a wide-reaching appeal. Who can resist an owl?

Unfortunately, the Little Owl has been suffering a decline in recent years, although the reasons for this are unclear. According to BTO statistics, its population fell by 46 per cent between 1967 and 2007 alone, but as a non-native species, it is unlikely that conservation efforts will be made to halt this decline.

Ring-necked Parakeet

This bright green member of the parrot family is steadily making inroads across the British landscape and is probably a less popular introduction. The Ring-necked Parakeet, which is also known as the Rose-ringed Parakeet, was popular as a pet and gradually established itself here after escaping from captivity. Its natural range encompasses a wide range of habitats, so this exotic-looking bird has no problems at all surviving in our cool climate. It's a sociable bird, so you're likely to encounter it in small flocks and its loud parrot-like squawk is unmistakeable. It thrives in an urban environment, feeding on seeds, nuts, fruits and berries, so municipal parks and suburban gardens make an ideal home. Its stronghold is in London and the Home Counties, though it is gradually spreading across the country and has been recorded as far afield as Dorset, Manchester and Liverpool. So if you have a garden with fruit trees and bird feeders, watch out! A gang of handsome green bandits with black neck rings and pink beaks could be launching a raiding party near you soon.

Ruddy Duck

The Ruddy Duck is a bird that has come and almost gone again from our freshwater lakes. It originates from the Americas and the drake is particularly handsome, with a rufous-chestnut body, black crown, white cheeks and a pale blue bill. As such, it was a popular addition to Sir Peter Scott's wildfowl collection at Slimbridge in Gloucestershire, from which it escaped in the 1950s and spread throughout Britain, thriving in the abundant habitat here. Meanwhile in Spain, however, its close relative, the White-headed Duck, was becoming endangered and urgent measures were being taken to protect and preserve the species. Then, hybrid Ruddy/White-headed Ducks were found in Spain and the prime suspect was the Ruddy Duck from Britain. In order to protect the purity of the White-headed Duck species, the controversial decision was taken to eradicate the Ruddy Duck from Britain and marksmen were employed to despatch the duck. There are now fewer than 150 individuals left in Britain, though if you are sharp-eyed enough, you may just spot one lurking on the edge of a quiet lake and keeping its head down.

Winners and Losers in the World of Birds

'Change is the only evidence of life', wrote Evelyn Waugh in *Brideshead Revisited*, and change is something that anyone who watches birds will notice for themselves.

You may, for example, notice a change in bird numbers. The BTO's Breeding Bird Survey has been carried out annually for the past 20 years and has tracked how different species have responded to changes in our landscape and our climate. Some species have benefitted; for example Little Egrets have shown an impressive 1,500 per cent increase over the past twenty years, and Common Buzzards are rising across the UK with an increase of over 44,000 per cent in some specific areas. Two red-listed species, Tree Sparrows and Lesser Redpolls, have both shown a slight increase in the latest survey. Other species have, sadly, gone into decline. Waders are on a downward slide, with Curlews and Redshanks decreasing by more than 40 per cent over the course of the 20-year survey, while Starlings, once the commonest bird being surveyed, has suffered a decrease of 51 per cent across the whole of the UK, and 70 per cent in Wales alone.

You may also notice a change in landscape use, particularly in farming practices which are under political pressure to achieve high productivity at low cost. This has led to the land being farmed much more intensively and mechanically, which has in turn had a detrimental effect on native birdlife such as Grey, Pied and Yellow Wagtails, Turtle Doves, Corn Buntings and Yellowhammers. At the same time, our growing human population is demanding more housing, more roads and more services, all of which are eating into our countryside. According to the Committee on Climate Change, 'Three-quarters of the remaining natural wildlife habitat is small and fragmented, making it more vulnerable to climate change. The proportion of our most important wildlife sites in good condition has fallen from 42 per cent to 37 per cent over the last decade. Sixty per cent of native species are experiencing long-term decline in populations.'

However, it is not all a story of doom and gloom. The past 20 years have also shown a huge increase in the number of people who go out to visit nature reserves and protected countryside to look for birds and other wildlife, and experience nature in general. According to a report by DEFRA (Department for Environment, Food and Rural Affairs) on the number of people choosing to spend their leisure time visiting a nature reserve run by the RSPB, The Wildlife Trusts or Natural England, the number of annual visits has increased consistently every year, while the National Trust, the UK's largest conservation organisation, reports a similar huge increase in the number of visits. This is all good news: the more people choose to visit reserves for pleasure, the more they will care about their wildlife and take steps to ensure its protection.

So, join the growing numbers of people who go out and watch birds. You will notice the changes in the seasons and the changes in bird plumages as the year goes by. You may notice changes in habitat and in the bird numbers you see over the years. But one thing will never change; the sheer pleasure you get from simply watching birds.

What are you waiting for? Get out there and experience for yourself the joy of birdwatching!

BIRDWATCHING JARGON

There are several words that loosely denote that a person is into birds… And there's a sort of hobby snobbery involved.
BILL ODDIE

Bins: most commonly used current short-form of binoculars; they have also been called binocs, binos and noccies in the past

Bird: a beautiful creature with feathers, two legs and two wings that can bring hours of pleasure when you watch them, or make a grown man cry when he 'dips' one of them (see 'To dip' below)

Bird guide: a knowledgeable person who takes you out to show you birds; there are a number of professionals who do this for a living

Bird tour: a short break or holiday to a place, region or country specifically to look for birds

Birder: someone who goes out somewhere specifically to look for birds; rarely seen without binoculars

Birding: actively seeking out interesting birds

Birdwatcher: someone who goes out for a walk and looks at birds if they happen to encounter them

Birdwatching: to watch birds wherever you encounter them; the subject of this book

Blocker: a very rare bird that you've seen and is on your list, but which is unlikely to be seen again for a long time

Call: the noise a bird makes to communicate, either to keep in contact with other birds, or to warn of a predator (the latter being often short and monosyllabic)

Chum: a mix of fish offal and food waste thrown overboard to attract seabirds on a pelagic trip

Crippler: any rare and spectacular bird that shows brilliantly

To dip: to miss seeing a bird which you were looking for

Dude: a birdwatcher who may have a lot of equipment but who doesn't really know all that much about birds (all the gear, but no idea!)

Field guide: an illustrated book of the different species of birds in a given geographical area, to assist in identification

A first: a first record of a species in a particular area, such as a country or county

To flush: to disturb a bird from where it has been sitting/standing/swimming

To grip someone off: to see a bird which another birder missed and tell them you've seen it

Gropper: slang name for a Grasshopper Warbler

Jizz: the overall impression of a bird given by its general shape, size, movement and behaviour, etc., rather than any particular feature. Experienced birders can often identify species, even with only fleeting or distant views, based on jizz alone

LBJ: abbreviation for Little Brown Job, a dull brown bird, of no particular species

LEO: abbreviation for Long-eared Owl

LRP: abbreviation for Little Ringed Plover

Lifer: your first-ever sighting of a bird species; an addition to your life list

List: a list of all species seen by a particular observer. It may be your life list, country list, garden list, office-window list, toilet list...

To lump: to reclassify two species that were previously thought to be separate as one single species of bird

Mega or Mega-tick: a very rare bird

Pelagic: a) adjective to describe an ocean seabird; b) a boat trip out to sea to look for ocean seabirds, may often involve rough weather and seasickness!

PG Tips: slang name for Pallas's Grasshopper Warbler

Plastic: adjective used to describe a bird that has escaped from captivity, as opposed to a genuinely wild bird

RB Fly: abbreviation for Red-breasted Flycatcher

SEO: abbreviation for Short-eared Owl; nothing to do with websites and search engines!

'Scope: short for telescope

Sibe: a vagrant bird from Siberia, e.g. Siberian Thrush

Song: the noise a bird, usually male, makes to declare its territory or to attract a mate; usually longer than a warning or contact call, and often melodic

Spot Red/Spot Shank: slang names for a Spotted Redshank

To split: to reclassify as two separate species or sub-species a bird that was previously thought to be one species

To string: to claim a dubious or suspect record of a bird, usually a rare species which is often only seen by one person, i.e. with the sighting not substantiated by any other birdwatchers

Stringer: someone who is known to make 'dodgy' claims about the birds they have seen – this is the worst reputation a birder or birdwatcher can have!

Tick: an addition to a personal list such as a year tick or a garden tick

Tart's tick: a rather derogative term to describe a relatively common species added to your list later than might be expected

Twitcher: someone who travels to see rare or scarce birds to add to their year list and/or life list

Under the belt/UTB: slang for having seen a bird and put it on your list

Yank: a North American vagrant species found in the UK, e.g. Lesser Yellowleg

FURTHER READING AND RESOURCES

Internet resources:

www.rspb.org.uk – reserves to visit, bird identification, local groups

www.birdguides.com – information about bird sightings in Britain and Ireland

www.fatbirder.com – information about birding worldwide, supplies and equipment

www.surfbirds.com – bird news and features from around the world, photos, bird lists

www.10000birds.com – information and commentary on birds, birding and conservation worldwide

www.birdwatching.co.uk – online magazine with bird news, equipment reviews, articles about birds and where to go birdwatching

www.birdwatch.co.uk – online magazine for keen birders, with rare bird news and articles on bird identification

www.birdsofbritain.co.uk – information about bird reserves and clubs, and features about birds

www.britishbirds.co.uk – online resource with features on bird identification and conservation, and the British Birds Rarities Committee Annual Report

www.birdid.no/bird/index.php – online resource for learning about bird ID and testing your skills

www.xeno-canto.org – listen to bird calls and songs from all around the world and share your own recordings

Apps:

Birdguides – alert on the latest unusual bird sightings across Britain and Ireland as they are reported

Birdguides: Birds of Northern Europe – field guide with ID images and information on the birds found across Northern Europe

BTO BirdTrack – an app to help you log your bird sightings as you are out and about birdwatching, and make a contribution to citizen science at the same time

Collins Field Guide – online version of the definitive field guide to the birds of Britain and Europe

Rare Bird Alert – instant alert on breaking news of rare and scarce birds throughout the UK

Books:

Field Guides

Collins Bird Guide: the most complete guide to the Birds of Britain and Europe

The Crossley ID Guide: Britain and Ireland

RSPB Pocket Guide to British Birds

Reading

Kingbird Highway by Kenn Kaufman – the tale of a young birder on his quest to set a new American birding record

The Big Year by Mark Obmascik – three birders go head-to-head to see the most birds in North America in a single year

The Big Twitch by Sean Dooley – Sean spends a whole year, and his inheritance from his parents, birding around Australia to set a new country record

The Urban Birder by David Lindo – David's life story as birds take him from growing up in 1970s east London to the bright lights of Hollywood

The Biggest Twitch by Alan Davies and Ruth Miller – the tale of a British couple (the authors of this book!) who give up everything to spend a year birding around the world to see more bird species than anyone else in a single year.

How to Be a Bad Birdwatcher by Simon Barnes – a humorous look at how birdwatching has framed his life

A Bird in the Bush by Stephen Moss – a social history of birdwatching

Fighting for Birds by Mark Avery – a vivid account of Mark's lifelong fight for bird conservation, both during his career within the RSPB and independently

Bill Oddie's Little Black Bird Book – the ultimate tongue-in-cheek guide to birdwatching

Birders: Tales of a Tribe by Mark Cocker – the story and history of the birding community in Britain

Tales of a Tabloid Twitcher by Stuart Winter – the tales behind the tabloid stories about birds and birdwatching by an *Express* journalist

While Flocks Last by Charlie Elder – the tale of Charlie's travels around Britain to see the bird species on the UK's Red List

The Birdwatcher's Yearbook by Buckingham Press – an excellent resource of reserves to visit, groups to join, guides, speakers and more

Magazines

Bird Watching Magazine – see website (p.154)

Birdwatch Magazine – see website (p.154)

British Birds – serious magazine for keen birders including reports from the British Birds Rarities Committee

The Birdwatcher's Year

Richard Williamson

THE BIRDWATCHER'S YEAR

Richard Williamson

ISBN: 978 1 84953 436 9   Hardback   £9.99

This charming and practical handbook is bursting with tips, facts and folklore to guide you through the birdwatching year. Find out how to identify birds by sight or song, everything you need to know about their behaviour, habitats and breeding and migration habits, and tips for encouraging birds into your garden. Also includes handy diary pages for making your own notes each month. A must-have for any eager birdwatcher.

If you're interested in finding out more about our books,
find us on Facebook at **Summersdale Publishers**
and follow us on Twitter at **@Summersdale**.

www.summersdale.com